LASER PIONEER INTERVIEWS

With an Introduction to Laser History by Jeff Hecht

A Lasers & Applications Staff Project

HIGH TECH PUBLICATIONS, INC.
Torrance, California

The Laser Pioneer Interviews originally appeared, in slightly different form, in 1985 issues of *Lasers & Applications* magazine. The Gordon Gould and Charles H. Townes interviews appeared in January, Robert N. Hall in February, Peter Sorokin in March, J. J. Ewing in April, Theodore H. Maiman in May, William Silfvast in June, C. Kumar N. Patel in July, John M. J. Madey in August, William Bridges in September, Ali Javan in October, Nicolaas Bloembergen in November, and Arthur L. Schawlow in December.

Carole Benoit: Publisher
Gerald R. Black: Associate Publisher
Thomas R. Farre: Editor
Terri Ledesma: Production Director
Randy Niwa: Art Director
Cheryl Miller: Advertising Manager
Karen Angell-Cole: Circulation Manager

621.366
LAS

ISBN: 0-936551-00-3

Manufactured in the United States of America.

First Edition December 1985.

CONTENTS

PREFACE

This book began as *Lasers & Applications'* commemoration of the 25th anniversary of the operation of the first laser in 1960. The magazine's editors sought out laser pioneers and asked them to recall how laser technology evolved, resulting in a thirteen-article series of Laser Pioneer Interviews that appeared in *L&A* throughout the year.

Considerations of space and time prevented us from including everyone who played an important role in developing new lasers, but we did try to sample the important areas. Those interviewed include three Nobel laureates: Charles H. Townes, Arthur L. Schawlow, and Nicolaas Bloembergen. We also interviewed Gordon Gould, whose laser patent claims have made him very controversial — and may also have helped stimulate interest in laser history.

The other interviewees helped develop the lasers that have become the cornerstones of our industry. Theodore Maiman built the first laser of any kind — the ruby laser. The others played pivotal roles in developing today's major commercial types, as well as one of the most promising lasers in today's laboratories — the free-electron laser.

This is not a scholarly or formal history — it is intended to let readers share the thoughts of the laser pioneers. Those seeking more details can consult the publications listed in the biliography at the back of the book.

Much more extensive interviews have been conducted by The Laser History Project, a scholarly effort jointly sponsored by the American Physical Society, the Laser Institute of America, the Optical Society of America, and the IEEE (Institute of Electrical and Electronics Engineers) Lasers and Electro-Optics Society, with the cooperation of the IEEE Center for the History of Electrical Engineering and the American Institute of Physics Center for History of Physics. This project has interviewed many laser pioneers not included in this book, and its archives on early laser research will be invaluable to future historians of science.

ACKNOWLEDGMENTS

This book was a team effort, and could never have become reality without the help of many creative and hard working people.

Carole Benoit, founder and publisher of *Lasers & Applications*, realized from the start the significance of the project and gave it her strong backing each step of the way.

Jeff Hecht's contribution was invaluable throughout—he conducted many of the interviews, wrote a fascinating historical introduction to laser history, and performed much of the work in helping to enrich the magazine series and convert it into book form.

Joan Bromberg, director of The Laser History Project, gave valuable advice and comments, and provided the bibliography that served as a starting point for this book's final chapter.

Also thanks to the many *Lasers & Applications*' staff members, past and present, who contributed to this project, including Holly Bigelow, Gerald Black, Lee Branst, James Cavuoto, Richard Cunningham, C. Breck Hitz, Tom Farre, Susan Lamping, Terri Ledesma, Michael Moretti, Randy Niwa, and Bill Rosas.

But we owe the most appreciation to the laser pioneers who gladly gave of their time to share their thoughts with us.

—The Editors
Lasers & Applications
November 15, 1985

AN INTRODUCTION TO LASER HISTORY

by Jeff Hecht

The laser is a quarter of a century old, but its roots go back long before physicists had devised the theoretical concepts of the laser or its predecessor, the microwave maser. The thread of ideas leading to the laser can be traced back to Albert Einstein's original proposal that the emission of radiation could be stimulated as well as spontaneous. However, the concept of stimulated emission remained primarily of academic interest until the 1950s, when Charles H. Townes conceived the idea of — and built — the maser.

The coming of the maser stimulated the emission of a wealth of ideas from physicists around the world, with the greatest concentration of work in the United States and another cluster of activity in the Soviet Union. Interest soon turned to extending the maser concept to shorter wavelengths, particularly to visible light. Theoretical studies spawned a race to build the first laser that eventually was won by Theodore H. Maiman on May 16, 1960.

More lasers followed — as did controversy over who deserved credit for what. This book cannot resolve all of the controversies, but it does let the pioneers in the field speak for themselves. This introduction is intended as an overview of laser history, to put the observations of the laser pioneers into context so the reader can understand how they fit into the evolution of the world of lasers.

Stimulated Emission

Albert Einstein originated the concept of stimulated emission of radiation in a 1916 paper.[1] Physicists had believed that a photon could interact in only two ways with an atom or molecule. A species could absorb the photon and jump to a higher energy level, or a species in an excited state could emit a photon spontaneously and drop to a lower energy level. Einstein suggested a third possibility, that in certain circumstances a photon could stimulate a species in an excited state to emit another photon with the same energy. This stimulation process could work only if the stimulating photon had the same energy as a transition from the excited state to a lower energy level.

Those conditions are restrictive, but not so severely restrictive as they might seem. Spontaneous emission from the excited-state species is a ready source of photons with the right energy to stimulate emission from other atoms or molecules in the same excited state. However, when matter is at thermodynamic equilibrium, stimulated emission is unlikely because the lower energy level of a transition has a higher population than the upper energy level. Thus a photon corresponding to the transition energy is likely to be absorbed by an atom or molecule in the lower energy state before it can stimulate emission from one in the upper energy level.

After World War I, German physicist Rudolf Walther Ladenburg became interested in spectroscopic theory. He devoted much of his attention to the theory of oscillator strengths and atomic absorption, including the phenomenon of dispersion, the change of refractive index with wavelength. His studies of absorption effects near resonances led to an analysis of negative dispersion — an effect caused by the stimulated emission postulated by Einstein. In 1928 he reported the first measurements confirming the existence of negative dispersion and stimulated emission, or what was at the time called "negative absorption."[2] He continued his studies for a few more years, measuring properties of neon gas after discharges were passed through it, but never raising discharge currents to high enough levels for "negative absorption" or stimulated emission to dominate.

Later observers, such as Arthur L. Schawlow, have suggested that physicists in the 1930s stopped short of producing net negative absorption because of an obsession with thermal equilibrium, which then was thought to be the normal condition of matter throughout the universe. Physicists in the 1930s apparently did not think they could produce an inverted population — the condition needed for stimulated emission to dominate, with higher-energy states having larger populations than ones with lower energy.

In 1940, Soviet physicist V.A. Fabrikant observed in his doctoral thesis that population inversion was necessary for "molecular amplification" — that is, domination by what we call stimulated emission. He added, "Such a situation has not yet been observed in a discharge even though such a ratio of populations is in principle attainable... Under such conditions we would obtain a radiation output greater than the incident radiation, and we could speak of a direct experimental demonstration of the existence of negative absorption."[3] Later in that decade, Willis E. Lamb Jr. and R. C. Retherford came tantalizingly close to stumbling upon population inversions. In their studies of the fine structure of hydrogen — which led to Lamb's 1955 Nobel Prize in Physics — they briefly examined the population of various energy levels in a discharge tube. They said it was possible that an "induced emission could be detected," but didn't try to do so themselves.[4]

The Maser

"Maser" is an acronym for "Microwave Amplification by the Stimulated Emission of Radiation." The concept grew out of research in microwave spectroscopy following World War II, much of it made possible by the use of war-surplus microwave equipment. The maser idea appears to have been conceived independently three times: by Charles H. Townes of Columbia University, by Joseph Weber at the University of Maryland, and by Alexander M. Prokhorov and Nikolai G. Basov at the Lebedev Physics Institute in Moscow. It was Townes, together with a postdoctoral assistant and a graduate student, who built the first maser.

Townes joined the faculty of Columbia University in 1948 after several years at Bell Laboratories, and became involved in microwave spectroscopy. His research at the Columbia Radiation Laboratory involved generating millimeter waves, but by the spring of 1951 he was growing impatient with the project's slow pace. Townes recalls that the maser idea came to him while he was attending a scientific conference in Washington DC. He awoke early and left his hotel room to avoid disturbing his roommate, Arthur L. Schawlow, who was a bachelor at the time and accustomed to sleeping late. Sitting on a park bench, he mulled over the problems of generating millimeter waves. He recalls:

"Perhaps it was the fresh morning air that made me suddenly see that this was possible. In a few minutes I sketched out and calculated requirements for a molecular-beam system to separate high-energy molecules from lower[-energy] ones and send them through a cavity which would contain the electromagnetic radiation to stimulate further emission from the molecules, thus providing feedback and continuous oscillation.''[5]

Townes took the idea back to Columbia, where he enlisted the help of postdoctoral fellow Herbert J. Zeiger and doctoral student James P. Gordon. They worked with a beam of ammonia molecules, seeking to extract excited molecules and thereby produce a population inversion. They directed the beam of excited molecules into a cavity resonant at the 24-gigahertz frequency of the ammonia transition. They hypothesized that spontaneous emission by some of the excited ammonia molecules would stimulate others to emit at that wavelength. The resonant cavity would aid in coupling the emitted radiation to the excited ammonia molecules, thereby maximizing the amplification produced.

After two years, Townes, Gordon and Zeiger had spent about $30,000 of a Joint Services grant but had not produced a maser. Despite the pessimism of some outsiders, they pressed on, and got the maser to work. It was later that the word "maser" was coined and quickly adopted by other researchers — although some cynics quipped that the acronym really stood for "Means of Acquiring Support for Expensive Research."[6] Townes, Gordon, and Zeiger reported their results in a 1954 *Physical Review* paper.[7]

Townes was not the only one thinking of stimulated emission of microwaves. Weber analyzed the prospects for obtaining stimulated emission from an inverted population shortly after joining the electrical engineering faculty at the University of Maryland in 1951. His suggested method of producing a population inversion never has been used, but he did realize that stimulated emission would produce coherent radiation. However, unlike Townes he considered only amplification by stimulated emission — not the oscillation which Townes produced in the first maser with resonant cavity. Weber's work was presented at a 1952 conference and published in 1953.[8]

Meanwhile, Prokhorov was leading a group of young physicists in research on molecular spectroscopy at Moscow's Lebedev Institute. He and Basov began to study the control of the population of various energy levels as a way to enhance the sensitivity of a molecular spectroscope. This led to a detailed study, published in October 1954,[9] of how molecules with different energy could be separated in a molecular-beam system, and how amplification could occur in a group of excited molecules. For his doctoral thesis Basov assembled the first Soviet maser—a few months after Townes' maser operated at Columbia. For their contributions, Basov, Prokhorov, and Townes shared the 1964 Nobel Prize in Physics for developing the "maser-laser principle."

The field of maser research grew rapidly and involved many of the men who later pioneered laser development. Prominent among them was Nicolaas Bloembergen, who published the first proposal for a three-level solidstate maser in 1956.[10] Another laser pioneer, Ali Javan, also worked on three-level maser concepts. That work, which both mention in their interviews in this book, was important in laying the groundwork for the laser. In his molecular-beam ammonia maser, Townes used a two-level scheme in which the excited molecules were physically separated from unexcited ones. This produced the population inversion needed to sustain stimulated emission because the only molecules present initially were in the excited state. However, a different arrangement was needed if the excited species could not be physically isolated, as, for example, in a solid.

The three-level concept requires an external energy source to produce a population inversion. The first three-level scheme was proposed by Basov and Prokhorov in a 1955 paper. They envisioned two approaches for a gas maser. In one, an external radiation source would excite molecules from a ground state to a higher energy level. Maser action would occur between that upper level and a slightly lower one which was not populated by the excitation method. The other method envisioned an excitation mechanism that depopulated the lowest of three energy levels. Thus the lowest level would have a lower population than one slightly higher in energy, and a population inversion leading to maser action would occur between those low-lying states.

Bloembergen's proposal for a three-level maser in a paramagnetic solidstate material was another significant step. Like the Soviets, he envisioned producing a population inversion by exciting the active species to a high-energy level. His proposal was far more specific however. It allowed for tunable-wavelength output because it relied on Zeeman levels, which can be tuned in energy by changing an external magnetic field. Bloembergen proposed some specific materials, and his concept offered the important practical advantages of wide enough bandwidth and continuously tunable amplification, while maintaining the low noise levels of a maser amplifier. He describes how the first three-level masers were developed in his interview.

The Birth Of Laser Theory

Even before masers began proliferating in the last half of the 1950s, a few physicists had begun looking at the prospects for amplifying stimulated emission at wavelengths much shorter than microwaves. In the Soviet Union, V.A. Fabrikant and his students filed a patent application dated June 18, 1951, titled "A method for the amplification of electromagnetic radiation (ultraviolet, visible, infrared and radio waves), distinguished by the fact that the amplified radiation is passed through a medium which, by means of auxiliary radiation or other means, generates excess concentration, in comparison with the equilibrium concentration of atoms, other particles, or systems at upper energy levels corresponding to excited states." However, that application is said not to have been accepted initially by the Soviet Patent Office, and was not published until 1959.[11] It appears to have had little direct effect on laser research, even in the Soviet Union.

In the United States, Robert H. Dicke developed the concepts of superradiance and what he called the "optical bomb." These involve producing an inverted population, which after the excitation pulse was over, would generate an intense burst of light by spontaneous radiation. Separately, he proposed the use of a Fabry-Perot interferometer as a resonant cavity in a patent titled "Molecular amplification and generation systems and methods," which he filed in 1956 and which was granted in 1958.[12]

Townes and Schawlow were the first to publish a detailed proposal for the construction of a laser, which at the time they called an "optical maser." Schawlow had been a postdoctoral fellow under Townes at Columbia until leaving to accept a post at Bell Labs in 1951. The two continued working together on a book on microwave spectroscopy, although not on masers. They also maintained close personal ties — they married sisters — and Townes consulted at Bell Labs. In 1957 both began thinking about the possibility of "infrared and optical masers," and after discussing the idea over lunch at Bell Labs, decided to work together on the idea. They spent several months on the problem, as described in their interviews. This led to their now famous paper "Infrared and Optical Masers," published in the December, 1958 *Physical Review*.[13]

That paper had a profound impact on American laser research. Preprints circulated at Bell Labs and Columbia before the journal came out, and formal publication was the starting gun for the great laser race that culminated in the completion of the first laser. Some of the laser pioneers recall its impact in their interviews in the following chapters. Not everyone realized its importance, however. Bell Labs attorneys did not think the idea worth patenting, and filed a patent application only after Townes insisted. That led to now-expired US Patent No. 2,929,922.

Meanwhile, similar ideas were running through the mind of a 37-year old Columbia graduate student, Gordon Gould. At the time, Gould was working on his doctoral thesis under Polykarp Kusch, who shared the 1955 Nobel Prize in Physics with Willis Lamb. Gould was hardly the prototype of the establishment scientist of the placid 1950s. He and his first wife joined a Marxist study group in the late 1940s, and though he left both the group and his wife, that background haunted him in the witch-hunting 1950s. He had begun taking graduate courses at Columbia while teaching at the City College of New York, and later became a full-time graduate student after he lost the teaching job because he refused to identify other members of the study group.

Gould wrote down his laser ideas — including a definition of LASER as "Light Amplification by the Stimulated Emission of Radiation" — in late 1957, and had them notarized in what he hoped was the first step in getting a patent. He continued working out his ideas in notebooks, but realized that he would have to leave Columbia to work on the laser. That led to a series of misadventures Gould describes in his interview.

The question of who really deserves credit for inventing the laser remains controversial. Townes and Schawlow have been widely honored by the scientific community for their work. There is no doubt that their *Physical Review* paper had a profound impact, and was the biggest single event triggering many research efforts that led to early lasers. Gould's notebooks and their offspring — his patent applications and proposals for research funding — had only minimal circulation and essentially no impact on most of the scientific world.

Gould has never achieved the professional eminence of Townes and Schawlow, both Nobel laureates in physics. Yet Gould has made other

Some rough calculations on the feasibility of a LASER: Light Amplification by Stimulated Emission of Radiation.

conceive a tube terminated by optically flat

partially reflecting parallel mirrors. The mirrors might be silvered or multilayer interference reflectors. The latter are loss less and may have an ~~arbitrarily~~ high reflectance depending on the number of layers. ~~a~~ a practical achievement is ∫ 98% in the visible for a 7-layer ~~↯~~ reflector. Flats with closer tolerance than 1/100 λ are not available so if a resonant system is desired, higher reflectance would not be useful. However for a nonresonant system, the 99.9% reflectance which are possible might be useful.

    consider a plane standing ~~wave~~ in the tube. There is the effect of a closed cavity; since the ~~↯~~ wavelength is small the diffraction and hence the lateral loss is negligable.

∫ O.S. Heavens, "Optical Properties of thin Solid Films" (Butterworths Scientific Publications. London. 1955). P.120.

Facsimile of the first page from Gordon Gould's notebook, notarized Nov. 13, 1957, on which he defines LASER as "Light Amplification by Stimulated Emission of Radiation."

contributions in laser physics and fiberoptics and cannot be lightly dismissed as a mere crackpot or plagarist. When two laser patents finally were issued to Gould in the late 1970s, as described later, some observers characterized him as an underdog, an outsider who unfairly had been denied credit for his work by the scientific establishment. Veteran laser researchers have differing opinions of the importance of Gould's contributions. Ultimately, the question of whether anyone stole anyone else's ideas nearly three decades ago is something that cannot be resolved here; the parties involved speak for themselves in the pages that follow.

Basov and Prokhorov also began exploring the possibilities of moving the maser principle to shorter wavelengths. In 1958, Prokhorov proposed a scheme for producing submillimeter maser action using rotational transitions of ammonia molecules, but his scheme for generating the population inversion was similar to that in Townes' original maser, calling for separation of excited molecules from those with lower-energy. The following year Basov published a proposal for producing a population inversion and stimulated emission by exciting a homogenous semiconductor with electrical pulses. The Soviet group continued studying prospects for semiconductor lasers and in 1961 proposed the *p-n* junction laser concept, which was later demonstrated.

The Great Laser Race

Publication of the Schawlow-Townes paper stimulated many efforts to build what were at the time called optical masers. At Columbia, Townes started working with two graduate students, Herman Z. Cummins and Isaac Abella, on a potassium-vapor laser, a scheme discussed in his paper. Schawlow, at Bell Labs, considered ruby as a laser material, but in 1959 dismissed it as unsuitable, a decision based on inadequate data which he admits in his interview was a poor one. Many initial efforts concentrated on materials whose energy-level structures were well-known from spectroscopic studies; ruby already had proved useful as a solidstate maser material.

Others also were working on optical masers. Ali Javan started on the helium-neon laser before the Schawlow-Townes article was published, as he describes in his interview. Peter Sorokin's interest in lasers was stimulated by seeing the Schawlow-Townes article in print.

Gould, who had left Columbia to work on the laser, soon interested his employer, TRG Inc., in the idea, and the company put together a proposal for laser development. The Department of Defense's Advanced Research Projects Agency was so excited by the idea that it gave TRG a contract for $1 million rather than the $300,000 originally requested. Unfortunately, the program was illfated from the start, as described in the Gould interview. Gould himself could not get security clearance to work on the program he had conceived.

The Ruby Laser Works

Meanwhile, on the West Coast, Theodore Maiman was trying to use his knowledge of ruby masers to make a laser at Hughes Research Laboratories in Malibu, Calif. As he relates in his interview, he forged ahead, working alone, despite assurances by Schawlow and others that ruby was unsuitable for a laser. Maiman knew better, but those statements led to frowns by management. By the time he succeeded in making the ruby laser work for the first time, on May 16, 1960, he was not supposed to be working on the program.

Maiman's success is undisputed, but almost immediately he ran into problems in reporting that success. Hughes' management reacted enthusiastically once the laser worked and sponsored a full-fledged press announcement in early July. However, the public-relations photographer commissioned to immortalize the first laser on film wasn't satisfied with it. He thought the device too small and insisted that Maiman pose with a bigger flashlamp and ruby rod. Today Hughes is still distributing those pictures, showing Maiman with what isn't really the first ruby laser.

A more serious problem came when Maiman submitted his paper for publication. The then-new *Physical Review Letters* summarily rejected it as "just another maser paper." The journal's founding editor, Samuel Goudsmit, a theoretician best known as the co-discoverer of electron spin, had grown tired of the glut of maser papers arriving at his office and decided they no longer merited rapid publication. Maiman hurriedly prepared a concise 300-word report which was immediately accepted by the British weekly *Nature*, and when efforts to convince Goudsmit of his error failed, *Nature* carried the first report of the laser on August 6, 1960.[14] Maiman later published a more detailed analysis in *Physical Review*.[15]

In their interviews, some of the laser pioneers recall when and

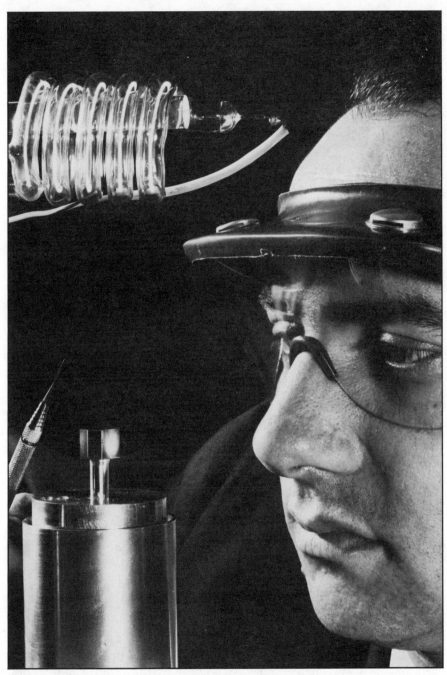

Early publicity photo released by Hughes showing Ted Maiman with the first laser — but which actually showed a larger flashlamp and ruby rod.

how they heard the news of Maiman's breakthrough. It was not long before other laboratories had duplicated Maiman's laser — some of them using the flashlamp shown in the press release photo, rather than the one Maiman actually used. Schawlow's group at Bell Labs was among the first to get one working, and soon afterwards they and another group reported laser action on slightly different lines in "dark" or "red" ruby, which has a higher concentration of chromium ions than the "pink" ruby Maiman used.[16]

The Great Laser Boom

Virtually everyone trying to build a laser had concentrated on elaborate schemes to generate an inverted population in the laser medium and to strengthen what they expected to be a weak signal. Maiman's laser was so elegant in its simplicity, and produced such unexpectedly powerful pulses, that it caused many researchers to rethink their approaches. In his interview Peter P. Sorokin recalls, after learning of Maiman's results, switching to a flashlamp-pumped rod design from a more elaborate approach involving a polished square of laser crystal. The first time Sorokin and Mirek Stevenson tried flashlamp-pumping a cryogenically cooled crystal of uranium-doped calcium fluoride, in November of 1960, it worked, and they had the second type of laser. [17]

Sorokin's laser emitted at 2.5 micrometers in the infrared, but because of factors such as the need for cryogenic cooling, it has never found many practical applications. However, it marked another important milestone — it was the first four-level laser.

The distinction between three- and four-level lasers is as significant as that between two- and three-level masers. At room temperature, ruby is a three-level laser, although at low temperatures it does become a four-level laser. Chromium atoms in the aluminum-oxide host are excited from the ground state to a higher energy level, from which they drop to the metastable upper level of the laser transition. The lower level of the laser transition is the ground state. Thus a majority of the chromium atoms in the crystal must be excited out of the ground state if the population in the upper laser level is to exceed that in the lower laser level. This had seemed a formidable problem to many early developers, but Maiman solved it by using intense light from a flashlamp to produce a transient population inversion that would generate a laser pulse.

The four-level approach that Sorokin pursued is, in general, a better one and is the basis of the neodymium-YAG laser and many others of practical importance. As in ruby, the atoms start in the ground state and are excited to a high-level state from which they decay to a metastable state that is the upper laser level. However, the lower laser level is not the ground state. Instead it is a level high enough above the ground state that it is largely unpopulated before laser operation begins. Thus it is much easier to achieve a population inversion because the lower level population is small to start with. This makes continuouswave laser action relatively simple to achieve in a four-level laser, while it is extremely hard to produce continuouswave beams from a three-level system.

Sorokin and Stevenson actually had considered two potential four-level solidstate lasers, with the second based on divalent samarium ions in calcium fluoride. A few weeks after they operated their uranium laser, they observed laser action at 708.5 nanometers at the edge of the red spectral region from the samarium laser.[18] Like the uranium laser, the samarium laser requires cryogenic cooling and has yet to find practical applications.

The First Gas Laser

The proliferation of solidstate lasers was ironic. Most early laser proposals involved gas lasers, including vapors of the Group 1 alkali metals. However, just before the end of 1960, the well-funded laser program at Bell Labs finally reported a triumph. Ali Javan, William R. Bennett Jr., and Donald R. Herriott at the Murray Hill, N.J. lab succeeded in building the first helium-neon laser.[19]

Javan became interested in optical lasers about the time he moved to Bell Labs from Columbia, where he had studied under Townes. He picked a gas discharge in a mixture of helium and neon as a particularly clean system and began the research he describes in his interview. The task was not an easy one because he lacked the tools available today. Mirror alignment was a particular problem because there were no off-the-shelf red helium-neon lasers to make parallel alignment of a pair of mirrors an easy job. After painstaking effort, Javan, Bennett, and Herriot finally got the laser to work at 4:20 pm on December 12, 1960.

Their helium-neon laser was not the familiar red HeNe of today. It operated on the 1.15-micrometer transition in the near infrared, which Javan's calculations had indicated was stronger than the red line. Javan says in his interview that the red line would not have worked then because gain is high only when the tube bore is so small in diameter that mirror alignment is difficult. In 1962, A.D. White and J.D. Rigden of Bell Labs obtained laser oscillation on the now-familiar 632.8-nanometer red line,[20] but by then Javan had moved to the Massachusetts Institute of Technology, where he remains on the faculty. A number of other lines have been discovered, but only recently have commercial versions become available on any but the 632.8-nanometer, and 1.15- and 3.39-micrometer transitions.

The advent of the helium-neon laser marked two crucial milestones. It was the first laser to emit a continuouswave beam — ruby and the other solidstate lasers demonstrated by Sorokin and Stevenson operated only in pulsed mode. This was possible because neon — the light-emitting species in the system — is a four-level laser. A continuouswave beam clearly has important advantages for many applications. The helium-neon laser also was the first gas laser. It not only showed that laser action was possible in a gas but also that it could be excited by a discharge and that it could rely on energy transfer between two different species.

The second gas laser remains more a fascinating footnote to laser history than a practical device. Paul Rabinowitz, S. Jacobs, and Gould at TRG succeeded in obtaining laser action from cesium vapor at 3.2 and 7.2 micrometers in the infrared by optical pumping with the 388.9-nanometer emission line of helium.[21] As Gould recalls in his interview, the TRG group had exhaustively studied a host of laser candidates before finally getting one to work. Their success certainly was not helped by security restrictions which limited how much researchers with clearances could tell Gould of their experimental progress. The system they finally demonstrated was similar to those proposed both by Gould and by Schawlow and Townes that relied on the near-coincidence of two spectral lines to create a population inversion in an alkali metal. However, such optically pumped systems never have proved practical.

The other laser demonstrated in 1961 was to have a much more important practical impact — the first neodymium laser, operated by L.F. Johnson and K. Nassau at Bell Labs.[22] Their first neodymium laser was in a calcium-tungstate host and emitted at the now-familiar 1.06-micrometer line. Neodymium is a four-level laser and has proved to be one of the most practical. Before long this laser was operated continuouswave at room temperature — an important breakthrough for solidstate lasers. Meanwhile, later in 1961, Elias Snitzer demonstrated the first neodymium-glass laser at American Optical.[23] Other researchers also investigated different hosts for the neodymium ion, but it was not until 1964 that laser action was demonstrated in what is now the standard neodymium host, yttrium-aluminum garnet (YAG), by J.E. Geusic, H.M. Marcos, and L.G. Van Uitert at Bell Labs.[24]

The First Commercial Lasers

It didn't take long for laboratories around the world to duplicate Maiman's ruby laser, but not all laboratories wanted to try building their own. The first commercial lasers began appearing about 1961 — large ruby types designed to deliver high-power pulses. Maiman left Hughes shortly after reporting the ruby laser to head a group developing commercial lasers at a short-lived company called Quantatron, in Santa Monica, Calif. When Quantatron ran out of money, Maiman took the laser group to form Korad. Funding came from Union Carbide under an agreement that gave them the right to acquire control of the company in five years. Korad soon became a major force in the laser market.

Other companies also built ruby lasers. One of the first in the market was Trion in Michigan, later to be acquired by Lear Seigler and renamed the Lear Seigler Laser Systems Center. Of the early ruby laser makers only Raytheon is still in the solidstate laser business. American Optical got an early start in the glass-laser field, but never effectively capitalized on its work. Its laser group eventually spun off as the independent Laser Inc., which later was acquired by Coherent Inc.

Meanwhile, other companies started working on gas lasers, with the initial emphasis on helium-neon. One was Spectra-Physics Inc., founded in 1961, which developed a commercial laser as a joint venture with Perkin-Elmer. Another was Optics Technology Inc., which fiberoptics pioneer Narinder Kapany founded in 1960, and which eventually became involved in producing helium-neon lasers.

Related Developments

As early lasers became available in more and more laboratories, new discoveries were made about both their properties and potential applications. In 1961, A.G. Fox and Tingye Li of Bell Labs published an extensive theoretical analysis of oscillation modes in laser resonators.[25] Although their study was completed before anyone had operated a continuouswave laser, it is considered a definitive work in the field. Peter A. Franken and colleagues reported production of the second harmonic of the red ruby laser line[26] — a discovery which stimulated the entire field of nonlinear optics as Nicolaas Bloembergen discusses in his interview. Robert W. Hellwarth and R. J. McClung at Hughes developed the Q switch and "giant-pulse" ruby lasers[27], which concentrated laser energy emission in a nanosecond-scale pulse with higher peak power than otherwise obtainable.

These results, in turn, helped enhance the usefulness of lasers. The study of laser oscillation modes led to improvements in resonator structure, particularly important for continuouswave low-gain lasers such as helium-neon. Nonlinear optics and harmonic generation made new wavelengths available. Q-switching allowed higher peak powers and shorter pulses — developments which helped make new applications such as laser rangefinding practical, as well as opening up new areas for research.

The pace of laser discoveries picked up in 1962 with the emergence of the red helium-neon laser and a few other types. That year also saw the discovery of an entirely new type of laser — the semiconductor diode or injection laser.

The Semiconductor Laser

Semiconductor physics was a hot topic in the late 1950s, and in retrospect it is not surprising that a number of people considered the possibility of producing a population inversion and maser or laser action in semiconductors. John von Neumann considered light amplification by stimulated emission in semiconductors in notes that he made in 1953, but which were not published until after his death.[28] On April 22, 1957, two Japanese researchers, Yasushi Watanabe and Jun-ichi Nishizawa, filed a patent application on a "semiconductor maser" based on amplification of recombination radiation produced by carrier injection in a semiconductor, and received Japanese patent 273217 in September, 1960. They proposed a completely enclosed resonant cavity, as used in the microwave region, using as an example recombination radiation near four micrometers in tellurium.

Probably the most extensive theoretical studies were done by Basov's group in Moscow, which made a number of proposals. In 1961 they proposed using *p-n* junctions in highly doped degenerate semiconductors,[29] the approach that eventually proved successful. Detailed analyses of the prospects for stimulated emission were made by Maurice G.A. Bernard and G. Duraffourg in France and independently by Basov's group.

Meanwhile, some experimental progress was being made. In early 1962, a Soviet group reported observing line-narrowing in the light emitted by gallium arsenide diodes operated at 77 K and with high current densities. R.J. Keyes and T.M. Quist of MIT Lincoln Laboratories in June reported obtaining incoherent emission from GaAs diodes with internal quantum efficiency they estimated at 85%. They and other semiconductor researchers described their work at the Solid State Device Research Conference in July, and that stimulated the excitement which Robert Hall describes in his interview.

The progress reports triggered a race to build the first semiconductor laser, a race which nearly ended in a photo finish. Hall attended the meeting and took his ideas for making a semiconductor laser back to General Electric's Research and Development Laboratories in Schenectady, N.Y. By September, he had a working diode laser.[30] Within a matter of days, Marshall I. Nathan's group at IBM Watson Research Center in Yorktown Heights, N.Y., followed with a paper received October 4 by *Applied Physics Letters*.[31] Later that month Keyes and Quist at Lincoln Labs[32] and Nick Holonyak at the General Electric Laboratory in Syracuse[33] also reported semiconductor lasers.

The four independent groups had come up with remarkably similar lasers. The first three used *p-n* junctions in gallium arsenide, with the material cooled to the 77 K temperature of liquid nitrogen and pumped with microsecond pulses of high current intensity. Holonyak worked in gallium arsenide-phosphide instead, producing output at 600 to 700 nanometers instead of in the 840-nanometer region where the GaAs lasers operated. All but the IBM group polished the ends of the semiconductor crystal to form laser cavities.

These first semiconductor lasers were only a beginning, and it would take years for more practical technology to evolve. They required high threshold currents — on the order of 10,000 amperes per square centimeter — and were amenable only to pulsed operation at low temperatures. These high currents reflected an inherent limitation of the early semiconductor laser designs, poor current confinement. These early lasers were "homojunction" lasers, in which the active layer was bounded on both sides by the same material. Current confinement was improved — and threshold current reduced — by the later development of the heterojunction or heterostructure diode laser. In these devices, the active layer was bounded by one or two layers of a semiconductor compound of different composition with higher bandgap. In a single-heterojunction laser, one such layer was adjacent to the active layer; in a double-heterojunction laser, the active layer is sandwiched between two layers of different composition. Single-heterojunction lasers provide adequate confinement to allow pulsed operation at room temperature. However, double-heterojunction lasers were the key development permitting continuouswave operation of diode lasers at room temperature, first demonstrated in 1970 at Bell Labs.

It took a few more years for continuouswave room-temperature diode lasers to reach the commercial market, but once they did they sparked a tremendous expansion in the semiconductor laser market that continues today. New structures confined the active layer not merely within a thin layer, but within a thin stripe, further reducing threshold currents and improving the quality of output and device lifetime. Mass production techniques were developed allowing high-quality diode lasers to be sold for just a few dollars each in quantity, for such applications as audiodisk players. Unit sales of diode lasers approached the million mark in 1984 and passed it in 1985.

The Gas Laser Boom

The first high-power gas laser, the 10.6-micrometer carbon-dioxide laser, was discovered by C. Kumar N. Patel at Bell Labs in 1963. As Patel recalls in his interview, many atomic gas lasers had been discovered by that time but all were limited in power. Bell Labs had built a monstrous helium-neon laser that generated only 150 milliwatts. Patel decided to take a different approach and look at molecular gases. Carbon-dioxide was his first choice, and after calculations indicated it should work well, he tried exciting the pure gas in an electrical discharge and got tens of milliwatts on the first shot.[34] Shortly afterward he realized that his initial dismissal of diatomic molecules had been premature, and he went back to try carbon monoxide, which also worked. Patel and others gradually refined the carbon-dioxide laser, first adding nitrogen to transfer the discharge excitation more efficiently to the carbon-dioxide molecules, then adding helium or water to de-excite the lower laser level in carbon dioxide and thereby increase laser output power. That work also led to other infrared molecular gas lasers that operated on similar principles.

The ion laser was the next major type to appear. The first in the family was a mercury-ion laser discovered in late 1963 by Earl Bell and Arnold Bloom at Spectra-Physics Inc.[35] They observed four lines, two in the visible and two in the infrared, from a pulsed gas discharge in a tube containing a 500-to-1 mixture of helium and mercury, with peak power to 40 watts. Efforts to duplicate and understand the HeHg laser were what led William Bridges to discover the rare-gas argon-ion laser, as described in his interview. The discovery was an accidental one that came from Bridges' efforts to understand the effects of the buffer gas on mercury-laser operation. He got a neon-mercury laser to work but ran into problems with an argon-mercury version. To make sure that the mirrors still were aligned, he pumped out the tube and refilled it with a helium-mercury mixture. It worked — and also produced a previously unobserved line at 488 nanometer, which Bridges discovered came from argon remaining in the tube. He then tried pure argon in a clean tube and observed laser action on the now-familiar visible lines of singly ionized argon.[36] Those lines were discovered independently by two other groups, Guy Convert's at CSF in France, and William Bennett and some of his graduate students at Yale University, where he had returned after spending a year at Bell Labs.

Bridges' work touched off a flurry of research on laser action in rare-gas ions, leading to demonstrations of krypton, xenon, and neon-ion lasers. Those early lasers all were pulsed, although the pulsed behavior of the lasers had hinted at the possibility of continuouswave laser action. As Bridges relates in his interview, Eugene Gordon demonstrated the first continuouswave ion lasers at Bell Labs[37] by shrinking the discharge bore to a millimeter in diameter, thus raising the current density to levels high enough to sustain continuous laser action.

The mercury-ion laser also led to the development of metal-vapor lasers, both in neutral and singly ionized species. It inspired Grant Fowles, a professor at the University of Utah, to begin a systematic search for metal-vapor lasers with graduate student William Silfvast, who describes the results in his interview. Their first successes were in early 1965 with zinc and cadmium.[38] They were concentrating on the metals and at first did not realize that the helium buffer gas they used to carry the discharge through the tube was essential for laser action on some transitions, including the 441.6-nanometer blue line used in today's commercial helium-cadmium lasers.

Initially metal-ion lasers operated only in pulsed mode. Silfvast made the first continuouswave HeCd laser in late 1967 shortly after he came to Bell Labs. While looking for a lead-ion laser at Utah, Silfvast and Fowles found an unexpectedly strong line in the red, which was particularly surprising because their mirrors had very high losses in the red. That lead line was the first in a series of neutral-atom lasers that include manganese and the copper-vapor laser, the latter recently having become a commercial product. Neutral metal-vapor lasers are self-terminating types which can only operate in short-pulse mode because the lower laser level fills more quickly than it can be depopulated. The Utah work was picked up by the TRG group, which demonstrated the first manganese-vapor laser. Members of that group, including Gordon Gould, reported the first copper-vapor laser in 1966.[39]

Chemical Lasers

The use of chemical reactions to produce a population inversion and drive a laser first was proposed by Canadian chemist J. Polanyi at a June 8, 1960 meeting of the Royal Society of Canada — during the time between Maiman's demonstration of the first laser and its public disclosure. He submitted a paper outlining the concept to *Physical Review Letters*, only to suffer the same type of summary rejection as Maiman; the paper eventually appeared in the *Journal of Chemical Physics*.[40]

Although much research was devoted to chemical lasers during the next few years, and conferences on the topic were held in Moscow and San Diego, it was not until 1965 that J.V.V. Kasper and George C. Pimentel demonstrated the first chemical laser, a 3.7-micrometer hydrogen-chloride laser, at the University of California in Berkeley.[41] Other demonstrations followed, eventually leading to continuouswave chemical lasers which could lase purely from the energy produced in a chemical reaction, without outside energy sources.

The Dye Laser

The use of organic dyes as the active media in lasers was first suggested in 1961, but there was little experimental effort to pursue the idea. One unsuccessful attempt by D.L. Stockman of General Electric may have discouraged others. Construction of the first operating dye laser is described by Peter Sorokin in his interview.

Sorokin and John Lankard had been working with organic dyes as saturable Q switches for ruby lasers, and Sorokin became interested in the spectral properties of the dyes. He first observed strong emission when illuminating chloro-aluminum phthalocyanine in ethyl alcohol with pulses from a ruby laser. When mirrors were aligned with the dye cell, he and Lankard saw infrared laser action in the dye so intense that it burned their photographic emulsion.[42] Sorokin's early-1966 results did not get wide attention quickly, and they were independently duplicated by Mary Spaeth and D.P. Bortfield and by Fritz P. Schaefer's group at the Max Planck Institute in West Germany. Sorokin next tried flashlamp pumping, which also worked.

In these experiments, the dye-laser wavelength was fixed at the peak of the dye's fluorescence spectrum. The first dye lasers emitted in the infrared, but by 1967 visible-wavelength laser action had been demonstrated. That year also marked the first operation of a tunable dye laser, by Bernard H. Soffer and B.B. McFarland at Hughes Research Labs.[43] By replacing one mirror in a laser-pumped dye laser with a diffraction grating which could be turned, they showed that the dye's natural broadband emission could be made narrower and tuned over a wide spectral range. This tunability has proved the most important property of dye lasers and has led to their wide use in research laboratories.

All of the early dye lasers were pulsed. It was not until later that Benjamin Snavely's group operated the first continuouswave dye laser, pumped by a continuouswave argon-ion laser. That type has become important in many high-precision spectroscopy experiments and also has been adapted for the production of ultrashort pulses by modelocking techniques.

High-power Lasers For Weapons

The 1960s saw increases not just in the number of lasers but also in the power levels available from lasers. Although some interest in high powers came from potential industrial users, the big push came from the Department of Defense, which saw the laser as a potential weapon. Gould recalls that military interest, evidenced by the security restrictions imposed on the TRG program, in his interview. In 1962, General Curtis LeMay made a prophetic suggestion that lasers might be used to defend against missile attack.[44] The general press also picked up the idea of laser weaponry and produced some wildly speculative articles. One of them, titled "The Incredible Laser," appeared in a Sunday newspaper supplement late in 1962. The talk of plans for laser cannons and other science-fictional weapons was too much for Arthur Schawlow, and soon a copy of the article appeared taped to his door at Stanford, along with a message reading "For credible lasers see inside."[45]

Development of the carbon-dioxide laser at first raised new hopes for powerful lasers, but some reached monstrous proportions. One 8.8-kilowatt carbon-dioxide laser achieved that power by passing a beam through a 750-foot path. Perhaps it was this laser that prompted the crack: "A laser big enough to inflict militarily significant damage wouldn't even have to work — just drop it on the enemy." Military-related research also led to the development of the TEA (transversely excited, atmospheric pressure) carbon-dioxide laser, in which a pulsed transverse electric discharge is applied to a carbon-dioxide gas mixture at near atmospheric pressure. The result is production of short, intense pulses from the gas. (Normal carbon-dioxide lasers are operated at much lower pressures so the gas can sustain a stable electric discharge.)

The breakthrough that led to serious Pentagon efforts to build high-energy laser weapons was the 1967 development of the gasdynamic carbon-dioxide laser. By the time this work emerged from under a veil of security classification in 1970, continuouswave power levels had reached 60 kilowatts. Soon all three armed services were testing powerful gasdynamic lasers to evaluate the potential of laser weapons on the battlefield, in the air, or at sea. Progress in building powerful chemical lasers and discharge-excited carbon-dioxide lasers also led to construction of large versions of those types in both the United States and the Soviet Union. Although laser-weapon programs remain in the news as part of the Reagan Administration's Strategic Defense Initiative, the tactical laser weapon programs begun in the early 1970s have largely faded from view because of the severe difficulty of building battle-ready lasers and getting laserbeams through the atmosphere.

The Excimer Laser

The excimer laser is the most commercially important new laser developed during the 1970s. The term "excimer" actually is a misnomer, coming from the origins of the device. Strictly speaking, an excimer is a complex of two atoms which is stable only in an electronically excited state, such as the xenon dimer, Xe_2^*. A similar complex made up of two different atoms, such as xenon chloride, XeCl, should be called an "exciplex." However, the borderlines became blurred in the laser world because true excimer lasers were developed first, followed by "excimer" lasers in which the active species were molecules made up of two different atoms.

The first proposal for using excimers as the active species in a laser dates from 1960, even before Maiman reported his laser, but it was not until 1970 that a group at the Lebedev Physics Institute — Basov, V.A. Danilychev, Yu. M. Popov and D.D. Khodkevich — reported the first excimer laser.[46] Their demonstration relied on electronbeam excitation of liquid xenon, obtaining extremely shortwavelength laser action at 176 nanometers. In 1972 another group reported the first xenon laser based on electronbeam excitation of xenon gas.[47] High-power output from xenon was reported shortly thereafter, stimulating a round of intense interest in the vacuum-ultraviolet xenon laser that eventually wound down with the realization that the laser would not be practical.

Interest in rare-gas halide excimers was triggered by a 1974 paper by Don Setser of Kansas State University, who reported emission near 300 nanometers from what appeared to be xenon fluoride. As excimer-laser pioneer J.J. Ewing relates in his interview, this created an intense round of interest in rare-gas halide molecules. A number of laboratories mixed rare gases and halides, excited the mixtures with electronbeams, and watched what happened. The initial results were observation of broad emission lines.

The first laser action was observed by Stuart Searles' group at the Naval Research Laboratory in Washington in early 1975.[48] Ironically, that first excimer laser was xenon bromide, a species which has found few uses because of its low efficiency. A couple of weeks later, Ewing and Charles Brau at Avco Everett Research Laboratory got xenon fluoride to lase and, a week after that, xenon chloride. They got krypton fluoride to lase at 249 nanometers soon afterwards, obtaining the brightest output they had yet seen.[49] Those initial efforts were obtained with electronbeam pumping, an approach that has been pursued to develop high-power excimer lasers. Later in 1975 Ralph Burnham and Nick Djeu at the Naval Research Laboratory modified an old pulsed Tachisto carbon-dioxide laser and showed that discharge pumping was possible, a development that led to the present generation of commercial excimer lasers.

The Free-Electron Laser

The other promising laser to emerge from mid-1970s research is the free-electron laser. Its origins can be traced to a 1951 proposal by Hans Motz, then at Stanford University, to produce millimeter waves by passing a beam of electrons through an array of magnets of alternating polarity. The spatial variations in the magnetic field of the undulator would cause the electrons to emit radiation, similar to synchrotron radiation, as they were bent by the magnetic fields. Motz demonstrated such emission in the millimeterwave region, using a linear accelerator that is still being used at Stanford in small-scale free-electron laser experiments.

The field lay largely dormant until John M.J. Madey became interested in the idea in the late 1960s. His first proposal for a free-electron laser was published in 1971[50], and about that time he began work at Stanford on demonstrating the concept, which he recalls in his interview. After several years of effort, in 1976 his group succeeded in demonstrating a free-electron laser amplifier.[51] They amplified the beam from an external carbon-dioxide laser by passing it and an electronbeam through a wiggler magnet. The following year they demonstrated the first actual free-electron laser oscillator, obtaining peak power of 7 kilowatts at 3.4 micrometers in the infrared.[52]

Madey was not completely alone in the field, however. While he was concentrating on infrared and visible free-electron lasers, a separate group at the Naval Research Laboratory and Columbia University was working on similar devices for much longer wavelengths in the millimeter and submillimeter regimes. There are important differences in the physics of the two regimes. In the shortwavelength regime, the dominant physics are those of individual high-energy electrons in a low-current beam. The research on longer-wavelength devices has concentrated on the "collective" regime, where the dominant physics are those of lower-energy but higher-current electronbeams. Following some demonstrations of amplification of spontaneous emission at 400 micrometers and 1.5 millimeters, the NRL and Columbia groups collaborated in demonstrating a collective-regime oscillator producing peak power of 1 megawatt at 400 micrometers.[53]

An experimental lull followed those successes as theorists analyzed the free-electron laser concept. Much effort went into devising tests for claims that the free-electron laser would be tunable in output wavelength and able to generate high powers efficiently. The next round of experiments included operation of a visible-wavelength storage-ring free-electron laser at the University of Paris-South in Orsay, demonstration of kilowatt power levels at the Los Alamos National Laboratory, and various tests of physical predictions by other groups. Although the experiments are not easy to perform, the tests so far have seemed successful.

The X-Ray Laser

One continuing aspect of laser research that cannot be covered in these public interviews is the x-ray laser. Proposals for x-ray lasers appeared soon after those for visible lasers, but actual demonstration of an x-ray laser proved exceptionally difficult. The 1970s were enlivened by a few false alarms, and toward the end of that decade some people began giving up. The most significant defector from the x-ray laser race was the Defense Advanced Research Projects Agency, which had supported research in the field for many years.

Ironically, it was shortly after support dried up that the first real signs of progress appeared. In 1980, Geoffrey Pert's group at the University of Hull in England reported observing laser gain at 18.2 nanometers in a carbon plasma. Gain was produced by vaporizing thin carbon fibers with intense infrared laser pulses. Recombination of the carbon plasma produced the 18.2-nanometer emission. A detailed analysis concluded that this represented gain produced by stimulated emission.[54] The wavelength is properly considered to lie in the extreme ultraviolet, and the Hull group used that label. However, because the wavelength was much shorter than any previous observation of stimulated emission, outsiders tended to apply the "x-ray" label. It should be noted that the Hull experiments did not demonstrate oscillation because of the unavailability of oscillator mirrors at such short wavelengths.

The most dramatic report of x-ray laser progress is one that has never been officially confirmed because the experiment remains highly classified. Quoting unnamed sources, *Aviation Week & Space Technology* in early 1981 reported that physicists from the Lawrence Livermore National Laboratory had produced x-ray laser pulses at a wavelength of 1.4 nanometers by pumping the laser medium with thermal x-rays from a nuclear explosion.[55] Although that article did not name the people involved in the experiment, other sources indicated it was based on a concept by Livermore physicists George Chapline and Lowell Wood, and that the experiment was directed by Livermore physicist Thomas A. Weaver. Reflecting security restrictions, those researchers have not publicly provided details or confirmation of their work.

In October 1984, Livermore publicly announced the demonstration of what it was careful to call the first "laboratory" x-ray laser. A group headed by Dennis Matthews and including Mordecai Rosen, theorist Peter Hagelstein, and Weaver, used intense pulses from the Novette fusion laser to generate plasmas from thin foils of selenium and yttrium. The selenium experiments produced emission lines at 20.6 and 20.9 nanometers, yttrium at 15.5 nanometers.[56] Not only did the yttrium experiments produce a shorter wavelength than the Hull demonstration, but they also produced much stronger amplification. At the same time, a group from the Princeton Plasma Physics Laboratory headed by Szymon Suckewer reported much stronger gain on the 18.2-nanometer carbon line than that observed at Hull.[57] The Princeton experiment also used a laser-produced plasma, although in this case a magnetic field helped confine the plasma.

The Past Is Not Forgotten

The laser field has not forgotten its origins after a quarter of a century. Some of the pioneers remain involved in laser applications or research and development, including Sorokin, Schawlow, Javan, Madey, and Bloembergen. Others now devote more of their attention to other areas. Townes' main activities now lie in astrophysics. Patel has become a research manager at Bell Labs. Hall is working on problems relating to very large scale integration of semiconductor electronics. Gould, who remains a consultant to Optelecom after his early-1985 retirement, is involved in investment as well as fiberoptics.

The pioneers have received many awards for their work, only a fraction of which are listed in the introductions to their interviews.

Two of the professional societies in the field have named awards after laser pioneers. The Optical Society of America established the annual Charles Hard Townes award in 1980 for "outstanding experimental or theoretical work, discovery, or invention in the field of quantum electronics." Its recipients include James Gordon and Herbert Zeiger (who worked on the maser with Townes) in 1981, Patel in 1982, and Q switch developer Robert W. Hellwarth in 1983. The Laser Institute of America established the annual Arthur L. Schawlow medal for laser applications in 1982, and awarded the first medal to Schawlow "for distinguished contribution to laser applications in science and education." The American Physical Society, LIA, OSA, the IEEE Lasers and Electro-Optics Society, the IEEE Center for the History of Electrical Engineering, and the American Institute of Physics Center for the History of Physics are cooperating on the Laser History Project, which is recording interviews with laser pioneers and compiling documents on the history of the laser.

Ironically, it may have been Gordon Gould's persistent efforts to secure a patent on the laser which helped heighten the laser field's sense of history. Gould's early claims were frustrated by a series of "interferences" with other patents. Although he continued to believe he had a right to patent coverage, he ran out of money and energy to continue to fight. In the early 1970s, however, he enlisted the aid of New York-based Refac Technology Development Corp. Refac's resources were enough to do the job, and Gould was issued two patents: U.S. Patent 4,053,845 issued August 16, 1977, on "Optically Pumped Laser Amplifiers," and US Patent 4,161,436 on July 17, 1979, on a broad range of laser applications. A few companies have purchased licenses on the patented technologies, a few others have been sued for patent infringement. Ownership of the patents has changed since they were issued, and 80% ownership, 64% of the licensing revenue, is now held by Patlex Corp., which paid Gould $2 million for half of his remaining interest.

The Gould patents have been controversial ever since they were issued. Only one infringement case has come to trial, against the small General Photonics Corp., which could afford to do little more than roll over and play dead in the face of the legal power assembled to defend the Gould patents. Much time, effort, and money also has been consumed by seemingly endless legal maneuvering, including reexamination of Gould's patent claims by the U.S. Patent Office.

At this writing, the outcome of the re-examination seems to be denial of the patent claims, but the whole process and the decision is being appealed. Enough money is involved to virtually assure that the legal battles will continue until every avenue of appeal is exhausted.

Frontiers Of Laser Research

The short-wavelength and high-power domains are two important frontiers of current laser research. Like the x-ray laser program, both are sponsored by the military — specifically the Reagan Administration's controversial Strategic Defense Initiative. Research in both fields is subject to security restrictions that make it hard to assess the state of the art.

The large amounts of money budgeted for SDI are in one sense good news for researchers interested in probing these frontiers. The SDI Organization's Office of Innovative Science and Technology has promised to support innovative concepts in areas such as short-wavelength lasers, and to keep at least some of the basic research unclassified. SDI's directed-energy program is giving heavy support to efforts to raise the powers available from free-electron and excimer lasers, and to devise chemical lasers with shorter wavelengths than the infrared hydrogen-fluoride types which to date hold the record for highest continuous laser power — 2 megawatts from MIRACL, the Mid-Infrared Advanced Chemical Laser at the High Energy Laser National Test Facility in White Sands, N.M.

However, the goals of SDI also demand that much of the program's energy and money go into laser engineering — the essential but unglamorous task of converting a one-of-a-kind laboratory demonstration piece into hardware that can be produced efficiently and economically and operated reliably in environments much less benign than the laboratory. This vital aspect of laser development is largely omitted in this series of interviews because the focus is on the pioneers who discovered new lasers and new effects.

Even if SDI succeeds, it will not account for all the frontiers of laser research. Fascinating work is being done on ultrashort pulses. By mid-1985, researchers at what is now AT&T Bell Laboratories had succeeded in compressing pulses of laser light down to a length of 8 femtoseconds — equivalent to less than four full wavelengths. A whole new family of solidstate vibronic lasers, with output that is tunable in wavelength, is emerging from the laboratory.

Impressive achievements are being made in the semiconductor laser field. The performance of semiconductor lasers is pushing continually beyond old limits, with new devices able to transmit signals of gigabits per second, maintain linewidth under 10 kilohertz, or combine the outputs of a monolithic array of laser stripes to generate a watt or more of continuous output. New concepts such as superlattices are being explored that may give even more flexibilty in creating semi-conductor lasers to meet specific requirements. Some of the last decade or two's "new" lasers are being engineered into the coming decade's new technology. More than two decades after the red helium-neon laser emerged from Bell Labs, companies have begun producing green, orange, and yellow versions, using lines discovered years ago that had never before been exploited. The laser's second quarter century may not be as exciting as its first, but it will not be dull.

REFERENCES

1. A. Einstein, *Mitt. Phys. Ges.*, Zurich, Vol. 16, No. 18, p. 47 (1916); an English translation appears in B. L. van der Waerden, ed., *Sources of Quantum Mechanics* (North-Holland, Amsterdam, 1967).
2. R. Ladenburg, *Phys. Z.* Vol. 48, p. 15 (1928).
3. V.A. Fabrikant, Thesis (1940), quoted in Mario Bertolotti, *Masers and Lasers: An Historical Approach* (Adam Hilger Ltd., Bristol, England, 1983).
4. W.E. Lamb Jr. and R.C. Retherford, *Physical Review* Vol. 72, p. 241 (1947); W.E. Lamb Jr. and R.C. Retherford, "Fine structure of the hydrogen atom, part I," *Physical Review* Vol. 79, p. 549 (1950).
5. Charles H. Townes, "The early days of laser research," *Laser Focus* Vol.14, No. 8, pp. 52-58 (Aug. 1978).
6. Mario Bertolotti, *Masers and Lasers: An Historical Approach* (Adam Hilger Ltd., Bristol, England, 1983), pp. 75-84
7. J.P. Gordon, H.J. Zeiger, and Charles H. Townes, "Molecular microwave oscillator and new hyperfine structure in the microwave spectrum of NH3," *Physical Review* Vol. 95, p. 282 (1954).
8. J. Weber, "Amplification of microwave radiation by substances not in thermal equilibrium," *Transactions IRE Professional Group on Electron Devices PGED-3* p.1 (Jun 1953).

9. N.G. Basov and A.M. Prokhorov, "3-level gas oscillator," *Zh. Eksp. Teor. Fiz (JETP)* Vol. 27, p. 431 (1954).

10. N. Bloembergen, "Proposal for a new type solid state maser," *Physical Review* Vol. 104, p. 324 (1956).

11. Bertolotti, *Masers and Lasers*, p. 115.

12. R.H. Dicke, US Patent 2,581,652, "Molecular amplification and generation systems and methods," issued Sept. 9, 1958.

13. Arthur L. Schawlow and Charles H. Townes, "Infrared and Optical Masers," *Physical Review* Vol. 112, p. 1940 (1958).

14. Theodore H. Maiman, "Stimulated optical radiation in ruby," *Nature* Vol. 187, p. 493 (Aug. 6, 1960).

15. Theodore H. Maiman, "Stimulated optical emission in fluorescent solids, Part I, Theoretical considerations," *Physical Review* Vol. 125, p.1145 (1961); Theodore H. Maiman, R.H. Hoskins, I.J. D'Haenens, C.K. Asawa, and V. Evtuhov, Part II, *Physical Review* Vol. 125, p. 1151 (1961).

16. Duplication of Maiman's "pink" ruby laser (with low chromium-ion concentration) was reported in R.J. Collins, D.F. Nelson, A.L. Schawlow, W. Bond, C.G.B. Garrett, and W. Kaiser, "Coherence, narrowing, directionality, and relaxation oscillations in the light emission from ruby," *Physical Review Letters* Vol. 5, p. 305 (1960); production of emission on different lines in "red" ruby with higher chromium concentration was reported in A.L. Schawlow and G.E. Devlin, "Simultaneous optical maser action in 2 ruby satellite lines," *Physical Review Letters* Vol. 6, p. 96 (1961) and by I. Weider and L.R. Sarles, *Physical Review Letters* Vol. 6, p. 95 (1961).

17. P.P. Sorokin and M.J. Stevenson, "Stimulated infrared emission from trivalent uranium," *Physical Review Letters* Vol. 5, p. 557 (1960).

18. P.P. Sorokin and M.J. Stevenson, in J.R. Singer ed., *Advances in Quantum Electronics* (Columbia University Press, New York, 1961) p. 65; P.P. Sorokin and M.J. Stevenson, "Solidstate optical maser using divalent samarium in calcium fluoride," *IBM Journal of Research and Development* Vol. 5, p. 56 (1961).

19. A. Javan, W.R. Bennett Jr., and D.R. Herriott, "Population inversion and continuous optical maser oscillation in a gas discharge containing a HeNe mixture," *Physical Review Letters* Vol. 6, p. 106 (1961).

20. A.D. White and J.D. Rigden, "Continuous gas maser operation in the visible," *Proceedings IRE* Vol. 50, p. 1697 (1962).

21. P. Rabinowitz, S. Jacobs, and G. Gould, "Continuous optically pumped Cs laser," *Applied Optics* Vol. 1, pp. 511-516 (1962).

22. L.F. Johnson and K. Nassau, *Proceedings IRE* Vol. 49, p. 1704 (1961).

23. E. Snitzer, "Optical maser action of Nd^{+3} in barium crown glass," *Physical Review Letters* Vol. 7, p. 444 (1961).

24. J.E. Geusic, H.M. Marcos, and L.G. Van Uitert, "Laser oscillations in Nd-doped yttrium aluminum, yttrium gallium, and gadolinium garnets," *Applied Physics Letters* Vol. 4, p. 182 (1964).

25. A.G. Fox and Tingye Li, "Resonant modes in a maser interferometer," *Bell System Technical Journal* Vol. 40, p. 453 (1961).

26. P.A. Franken, A.E. Hill, C.W. Peters, and G. Weinreich, *Physical Review Letters* Vol. 7, p. 118 (1961).

27. R.W. Hellwarth, in J.R. Singer ed., *Advances in Quantum Electronics*, (Columbia University Press, New York, 1961) pp. 334-341; R.J. McClung and R.W. Hellwarth, *Journal of Applied Physics* Vol. 33, p. 828 (1962).

28. J. von Neumann, *Collected Works* Vol. 5 (Pergamon, London, 1963), p. 420.

29. N.G. Basov, O.N. Kronkhin, and Yu. M. Popov, *JETP* Vol. 10, p. 1879 (1961).

30. R.N. Hall, G.E. Fenner, J.D. Kingsley, T.J. Soltys, and R.O. Carlson, "Coherent light emission from GaAs junctions," *Physical Review Letters* Vol. 9, p. 366 (1962).

31. M.I. Nathan, W.P. Dumke, G. Burns, F.H. Dill Jr., and G. Lasher, "Stimulated emission of radiation from GaAs p-n junction," *Applied Physics Letters* Vol. 1, p. 62 (1962).

32. R.J. Keyes and T.M. Quist, *Proceedings IRE* Vol. 50, p. 1822 (1962); T.M. Quist, R.H. Rediker, R.J. Keyes, W.E. Krag, B. Lax, A.L. McWhorter, and H.J. Zeiger, "Semiconductor maser of GaAs," *Applied Physics Letters* Vol. 1, p. 91 (1962).

33. N. Holonyak Jr. and S.F. Bevacqua, *Applied Physics Letters* Vol. 1, p. 82 (1962).

34. Detailed results were published in C.K.N. Patel, "Continuous-wave laser action on vibrational-rotational transitions of CO_2," *Physical Review* A136, p. 1187 (1964).

35. W.E. Bell, "Visible laser transitions in Hg⁺" *Applied Physics Letters* Vol. 4, pp. 34-35 (1964).

36. W.B. Bridges, "Laser oscillation in singly ionized argon in the visible spectrum," *Applied Physics Letters* Vol. 4, pp. 128-130 (1964); erratum: *Applied Physics Letters* Vol. 5, p. 39 (1964).

37. E.I. Gordon and E.P. Labuda, *Bell System Technical Journal* Vol. 43, p. 1827 (1964).

38. William T. Silfvast, G.R. Fowles, and B.D. Hopkins, "Laser action in singly ionized Ge, Sn, Pb, In, Cd, and Zn," *Applied Physics Letters* Vol. 8, pp.318-319 (1966).

39. W.T. Walter, N. Solimene, M. Plitch, and G. Gould, "Efficient pulsed gas discharge lasers," *IEEE Journal of Quantum Electronics* QE-2, pp. 474-479 (1966).

40. J.C. Polanyi, *Journal of Chemical Physics* Vol. 34, p. 347 (1961).

41. J.V.V. Kasper and G.C. Pimentel, "HCl chemical laser," *Physical Review Letters* Vol. 14, p. 352 (1965).

42. P.P. Sorokin and J.R. Lankard, *IBM Journal of Research and Development* Vol. 10, p. 162 (1966).

43. B.H. Soffer and B.B. McFarland, *Applied Physics Letters* Vol. 10, p. 266 (1967).

44. Quoted in Paul B. Stares, *Space Weapons and US Strategy: Origins & Development* (Croom Helm, London, 1985) p. 111.

45. Jeff Hecht, *Beam Weapons: The Next Arms Race* (Plenum, New York, 1984) p. 26.

46. N.G. Basov, V.A. Danilychev, Yu. M. Popov, and D.D. Khodkevich, *JETP Letters* Vol. 12, p. 329 (1970).

47. H.A. Koehler, H.A. Ferderber, D.L. Redhead, and T.J. Ebert, *Applied Physics Letters* Vol. 21, p. 198 (1973).

48. S.K. Searles and G.A. Hart, "Stimulated emission at 281.8 nm from XeBr, *Applied Physics Letters* Vol. 27, p. 243 (1975).

49. C.A. Brau and J.J. Ewing, "354-nm laser action on XeF," *Applied Physics Letters* Vol. 27, p. 62 (1975); J.J. Ewing and C.A. Brau, "Laser action on the $^2\Sigma^+_{1/2} \to {}^2\Sigma^+_{1/2}$ bands of KrF and XeCl," *Applied Physics Letters* Vol. 27, p. 350 (1975).

50. J.M.J. Madey, "Stimulated emission of bremsstrahlung in a periodic magnetic field," *Journal of Applied Physics* Vol. 42, p. 1906 (1971).

51. L.R. Elias, W.M. Fairbank, J.M.J. Madey, H.A. Schwettman, and T.I. Smith, "Observation of stimulated emission of radiation

by relativisitic electrons in a spatially periodic transverse magnetic field," *Physical Review Letters* Vol. 36, p. 717 (1976).

52. D.A.G. Deacon, L.R. Elias, J.M.J. Madey, G.J. Ramian, H.A. Schwettman, and T.I. Smith, "First operation of a free-electron laser," *Physical Review Letters* Vol. 38, p. 892 (1977).

53. D.B. McDermott, T.C. Marshall, S.P. Schlesinger, R.K. Parker, and V.L. Granatstein, "High-power free-electron laser based on stimulated Raman backscattering," *Physical Review Letters* Vol. 41, p. 1368 (1978).

54. D. Jacoby, G.J. Pert, S.A. Ramsden, L.D. Shorrock, and G.J. Tallents, "Observation of gain in a possible extreme ultraviolet lasing system," *Optics Communications* Vol. 37, No. 3, pp. 193-196 (May 1981); D. Jacoby, G.J. Pert, L.D. Shorrock, and G.J. Tallents, "Observation of gain in the extreme ultraviolet," *Journal of Physics B: Atomic and Molecular Physics* Vol. 15, pp. 3557-3580 (1982).

55. Clarence A. Robinson Jr., "Advance made on high-energy laser," *Aviation Week & Space Technology* Feb. 23, 1981, pp. 25-27.

56. Dennis L. Matthews, paper presented at annual meeting of Optical Society of America, October, 1984, San Diego; M.V. Rosen et. al. *Physical Review Letters* Vol. 54, p. 106 (1985); D.L. Matthews et. al. *Physical Review Letters* Vol. 54, p. 110 (1985).

57. S. Suckewer, C. Keane, H. Milchberg, C.H. Skinner, and D. Voorhees, "Short review of recent soft x-ray laser development experiments at PPL," *Bulletin of American Physical Society*, Oct. 1984, p. 1211; S. Suckewer, C.H. Skinner, H. Mikhberg, C. Keane, D. Voorhees, "Amplification of stimulated soft x-ray emission in a confined plasma column," *Physical Review Letters* Vol. 55, pp. 1753-1756, Oct. 21, 1985.

CHARLES H. TOWNES

"Infrared and Optical Masers"

A native of Greenville, S.C., Charles H. Townes received two bachelor's degrees from Furman University in 1935, one in modern languages, the other in physics. He earned his doctorate in physics from the California Institute of Technology in 1939. Townes spent the war years at Bell Telephone Laboratories working on radar systems, including one in the then-unexplored 24-gigahertz range. This interested him in microwave spectroscopy, a field he continued to study after joining the Columbia University physics faculty in 1948.

He conceived of the maser concept in 1951, and worked with postdoctoral fellow Herbert J. Zeiger and doctoral candidate James P. Gordon on experiments that led to demonstration of the first maser in late 1953. Later, he and Arthur L. Schawlow collaborated on developing the theory of "infrared and optical masers," or lasers. Townes was provost of the Massachusetts Institute of Technology from 1961 to 1966, and in 1967 was named university professor of physics at the University of California, Berkeley, where his research includes astronomical masers.

In 1964, Townes shared the Nobel Prize in physics with Nikolai G. Basov and Alexander M. Prokhorov of the Lebedev Physics Institute for "fundamental work in the field of quantum electronics which has led to the construction of oscillators and amplifiers based on the maser-laser principle." Other honors include the 1958 Morris R. Liebmann award from the Institute of Radio Engineers (now IEEE), the National Academy of Sciences' Comstock Award in 1959, the Optical Society of America's C.E.K. Mees award in 1968, the NASA Distinguished Public Service Medal in 1969, foreign membership in the Royal Society of London, election to the National Inventor's Hall of Fame in 1976, the Neils Bohr International Gold Medal in 1979, and the National Medal of Science in 1982.

C. Breck Hitz conducted this interview on September 27, 1984 at the University of California, Berkeley.

L&A: What were you doing before the maser was invented?

Townes: I was working on microwave spectroscopy, studying the interaction between microwaves and molecules. This involved looking in detail at the spectra and interpreting them in terms of molecular structure and nuclear moments. It was atomic and molecular physics that I was doing, using microwaves as the spectroscopic tool.

L&A: It all developed from radar?

Townes: In a sense it did. During World War II, I worked on radar bombing systems, and became quite familiar with radar. In fact, one of the systems I worked on was at 1.25 centimeters, a system the Air Force wanted badly. But I believed that that wavelength was likely to be absorbed by water vapor. I tried to persuade people that it wasn't going to work, but I was too young and the decision had been made. Well, they went ahead and put it in the field, and it had no range because of water-vapor absorption. So all the equipment was junked.

In the meantime, I thought a good deal about interactions between molecules and microwaves and realized that it provided the possibility of some very powerful spectroscopy. I thought that it might be possible to do things that nobody had been able to do before, using very high spectral resolution. We had the equipment, which was very cheap— it was sold in the streets of New York almost as junk! And so, in that sense, the microwave spectroscopy work did derive from radar. We used that surplus 1.25-cm equipment for a long time in microwave spectroscopy.

That was about as short a wavelength as could be reasonably used at that time. Now, it's well known that molecular interactions with microwaves become stronger at shorter wavelengths. So I was very eager to get down into the millimeter or even submillimeter range. The primary object of the work that led to the maser was to get shorter wavelengths so we could do better spectroscopy in a new spectral region.

L&A: Was it understood then that a population inversion was relatively easy to create?

Charles H. Townes

Townes: Well, yes and no. Pound and Purcell had inverted populations at Harvard, and that was known to physicists in the business. It was less well known to engineers, I guess.

But the effects that Pound and Purcell were observing were very weak, and nobody saw any practical application for them. I had thought about stimulated emission from time to time, as had other physicists, but had never done an experiment on it because it seemed too weak. I think it was not that people felt they couldn't get inverted populations. What wasn't realized was that the effects could be made big enough to give significant amplification. And nobody recognized the possible importance of oscillation and large amplification if they could be achieved. Nobody had thought of feedback. Putting a rather large population inversion in a resonant cavity with feedback is really what made the maser possible.

L&A: Fabry-Perots had been around for a long time, of course.

Townes: Well, if you look back on it now, there's essentially nothing that wasn't known at that time by somebody. That's characteristic of many developments in science or technology. The breakthrough occurs when somebody realizes that a combination works in a way that no one had thought of before. Each individual element was known, and, if they had thought about it, yes, they would have realized it could work. But nobody did.

So, while I understood that there was nothing particularly difficult about producing some population inversion, I didn't realize initially the potential it had as a radiation source. When I was at Bell Laboratories, I wrote a little essay for my bosses, trying to persuade them that microwave spectroscopy was of some importance. But I said that while molecules can produce radiation, the radiation would be weak, like blackbody radiation.

Now, that's good thermodynamics for thermal equilibrium. But then came the sudden realization that, wait—one doesn't have to use thermal equilibrium. If you don't use thermal equilibrium, then you can have inversion of population and (in principle) the radiation intensity can become enormous. Further, a resonant cavity could help achieve this.

L&A: After the invention of the maser, did you immediately begin work on a visible maser—a laser? Or were there other projects that might have diverted you?

Townes: Well, as a matter of fact, at that point I took a sabbatical.

Townes with the second ammonia-beam maser.

I had just finished, with Art Schawlow, a book on microwave spectroscopy, and I'd also just finished a three-year term as chairman of the department of physics at Columbia University. I felt that microwave spectroscopy for physics was nearing completion. There was still a great deal of interesting chemistry that could be done with microwave spectroscopy, but the interesting things for physicists had been pretty well examined. It was time to think about what I wanted to do next.

So I cruised Europe and Asia—particularly France and Japan—for 15 months. I was looking at other fields, thinking about what I wanted to do and trying to decide whether to give up microwave spectroscopy. For a long time I'd been interested in radio astronomy, so I looked pretty hard at that field. But I was also thinking a lot about the maser. In France, one of my former students was working on a paramagnetic resonance with a long relaxation time. I realized that it could make a very good amplifier. We worked on that for a few months, the remainder of my time in Paris, and wrote a paper about it.

Then I left for Japan. Although I was still trying to decide what I should do next, the scientific work I did there was on the theory of amplification by stimulated emission. I worked with a couple of Japanese physicists, friends of mine there, in examining the discreteness of the resulting radiation field.

By the time I got home, I was convinced that maser development —later to be named quantum electronics—was the field I should continue with.

L&A: Did you ever try to build a visible maser?

Townes: Yes, I did, at Columbia. After Art Schawlow and I wrote the original paper, which was published in 1958, I started to build a laser. We were looking at the alkalis, particularly cesium and potassium, essentially what we discussed in our paper. We planned pumping with other atoms, but basically we were trying to get an inverted population in one of the alkalis. One of my students and I, and later another associate, worked on it for about a year. I must have started in the fall of 1958, a few months before the paper came out.

But in the fall of 1959, I accepted a job in Washington. There weren't many scientists in Washington at that time. I felt it was important for scientists to tackle some of the problems like arms control, space work, and other technical issues that the country was struggling over at that time. That was shortly after the Sputnik had gone up, you see. President Eisenhower had initiated the President's Science Advisory Committee, and there was a great scramble to get scientists more involved in questions of national policy.

I was urged to go down to Washington—I felt a duty to go—so I did. I still continued to come up to Columbia on Saturdays, to take care of my students and continue some of my laser work. But of course it was a pretty small effort so far as I was personally concerned.

L&A: Were you surprised when the Maiman laser first worked? Wasn't there a lot of doubt that a three level system like ruby could work at all?

Townes: I don't really think there was any doubt about a three level system in general. That's straightforward physics. Art Schawlow had also proposed a ruby system which was to work later, but Maiman's system was different and did involve levels which seemed more doubtful. I would have to say I didn't really expect making lasers to be as easy as it turned out to be. It seemed to me that people had done so much spectroscopy in the visible region, with gas discharges and other radiation sources, that somebody would have just stumbled into it if the laser were easy. So I felt it could not be easy, and everything must be planned carefully.

Art Schawlow and I based our specific calculations on the alkalis because the physics was well enough known that you could be sure a laser would work, if you did it right. That didn't mean it was the easiest case, but it was a case where we knew enough of the physics and enough of the parameters that it just had to work.

My style of physics has always been to think through a problem theoretically, analyze it, and then do an experiment which *has* to work. You analyze and duplicate the theoretical conditions in the laboratory until you beat the problem into submission, you see.

Now, there are other ways of doing physics; you can just try something and see whether it works. If it doesn't work, then go try something else. And that's a viable way of doing physics, too, but not the way I normally approach problems.

L&A: Was Maiman's laser an example of this second kind of physics?

Townes: Well, maybe. Certainly all the parameters weren't well known. On the other hand, Maiman had been measuring properties of ruby. He had looked at some time constants and relaxation properties. But one couldn't have known for sure beforehand that it would work. Of course, it did, greatly to his credit. But as it turns out, almost everything works if you hit it hard enough.

L&A: Has the alkali laser ever been "beaten into submission?" I mean, has the laser you and Schawlow originally proposed ever lased?

Townes: Yes, the cesium system has been made to lase. And I think most of the things we worked out in the 1958 paper surely can be done. There's just no longer great interest in the alkalis. There are so many better systems now, that just making another laser work isn't of great interest.

L&A: What role did the Russians, Prokhorov and Basov, have in all this? Were the US and Russian efforts independent?

Townes: Basov and Prokhorov made an independent proposal. It's hard to know just when they learned of my own work. I first met Prokhorov at a meeting of the Faraday Society in Cambridge, England, in the spring of 1955. He was already familiar with our 1954 publication on the maser. But I knew nothing about their maser work. Prokhorov showed up at the meeting and talked about the possibility of making a maser. I was giving a paper on something else, but of course I was quick to comment on our working maser. We had a good chance to talk and compare notes. It was rare at that time to have a good occasion for meeting Russian scientists, and it was a privilege to have a chance to discuss things with Prokhorov. I knew some of his work in microwave spectroscopy from his published papers, but I had never met him.

I didn't know anything about Prokhorov and Basov's maser work prior to the 1955 meeting, although they had submitted a paper for publication in early 1954, and to what extent they knew about our work by then is hard to say. In addition to our 1954 publication, I had discussed our maser work in Japan in 1953, given a paper about the newly operating maser to the American Physical Society in the spring of '54, and we had written about the maser as early as the December, 1951 Columbia University progress report. Now, those reports were generally circulated only to a list of about 100 labs in the US and Europe, and normally didn't go to the Soviet Union. But it turned out they had always been put on the open shelves at least in the Harvard library. Lots of visitors had also been through our lab, so it's hard to know now how much got around.

But my firm belief is that the initial paper of Basov and Prokhorov on the maser was independent and disconnected from ours.

Townes with an early laser (circa 1961).

L&A: Who are the other unsung heroes?

Townes: Well, I think Javan, Bennett, and Herriott were a very important group. They developed the helium-neon laser, which is now the most common of all lasers. Their system involved quite different ideas from Maiman's, and if Maiman hadn't made the first laser, they would have shortly later. They may be somewhat overlooked for not being the very first.

Actually, they weren't second, either, although they're sometimes credited with that. They were third. There's another group that's also overlooked, Sorokin and Stevenson, of IBM. They made two lasers from rare-earth salts, somewhat along the lines of Maiman's, and I bet you haven't heard of that.

L&A: I've always thought that HeNe was the second laser. A rare-earth salt? Was it optically pumped?

Townes: Yes; those were the second and the third lasers. They have had no substantial use. After that came the dark ruby lasers of Schawlow and others, and then Javan, Bennett, and Herriott with the HeNe laser. But I think this issue of exact time sequence is sometimes overemphasized. The role a contribution plays in the development of a field is perhaps more important.

There's another person whom you might consider unsung, a Russian scientist I have never met named Fabrikant. I don't think he's ever been out of the Soviet Union. He started in the 1940s, doing a thesis on inversion of population; as far as I know his ideas were original. He must have worked on the idea rather hard for a few years, though not much happened as a result. He applied for a patent in 1951, and when it was issued in 1959 it had a number of advanced ideas. But I just don't know enough about the Soviet patent systems to know what that means. I'm told that the Russian patent office allows quite a lot of changes between the time a patent application is filed and the time the patent is issued. Hence, what's published isn't necessarily what was initiated. Nonetheless, it's pretty clear that Fabrikant initiated something in the field in 1951. That was pretty early, and I think he deserves some credit.

L&A: Did Fabrikant interact with Prokhorov and Basov?

Townes: I don't know. Basov and Prokhorov, of course, deserve a good deal of credit, as do Maiman, Bloembergen, and Schawlow. But these people are generally well recognized for their contributions.

L&A: What future do you see for laser isotope separation and inertial confinement fusion?

Townes: Well, lasers work very well in isotope separation. But I don't think that's going to change the world in a big way, it just makes the process cheaper. The only way laser isotope separation might really change the world is not particularly cheering; if it becomes so efficient that the production of militarily important isotopes is cheap and easy, then that's not so good. But it won't be all that cheap in any case; it'll just be cheaper than current methods.

As for fusion, yes, I'm convinced now that lasers represent one way of making fusion work. Whether that'll be the best way, I don't know. And whether, overall, the cost of building a plant that would produce fusion is going to be economical by comparison with other energy sources, I don't know. It's quite possible, but I think 20 years is the soonest that anything like that could be in use, and I believe it is likely to take longer than that.

L&A: What about laser weapons and the strategic defense initiative?

Townes: I don't think anybody who has looked at it closely believes that optical lasers will solve the problem of defense against missiles. Maybe one can find somebody who believes they will, but the people I know don't believe it, nor do I.

X-ray lasers probably have some chance, but I don't really expect that they will work in the sense that they can overcome likely improvements in the offense. The trouble with the SDI program is not that we can't build fantastic things and shoot down almost any number of missiles, but rather that as one learns how to do that, the offense will learn too. And the offense seems likely to always have an advantage if offensive weapons are pursued as vigorously as defensive ones. So I don't think there's any way the defense can overwhelm the offense, unless further developments on the offensive side are neglected.

L&A: Of course, that's not the way the world works.

Townes: That's not the way the world works, and we have to allow for that. For that reason I'm not very optimistic about the SDI program really providing us with any reasonably complete defensive protection.

Now that doesn't mean it's a useless program. You can stop some weapons, and if it's cheap enough to do that, then obviously that's useful, even though it doesn't solve the complete problem. There are intermediate possibilities that one can consider. But I am pessimistic about the possibility of any really complete defense.

L&A: As you look back over the last 25 years, what about lasers has surprised you most?

Townes: Well there are several things that surprised me. One is how easy lasers really are to make. Considering the amount of work that had been done on optical spectroscopy before the 1940's, I'm still surprised somebody didn't make one accidentally.

Another thing that turned out to be better than I expected is the use of the lasers to excite other lasers. By using lasers to excite lasers, you can do things which otherwise seem very difficult, like building tunable dye lasers, or making x-ray lasers. I think dye lasers are really spectacular, providing the remarkable amount of tuning one can get in the optical region. I had initially thought it would be very difficult. I thought you would have to be very inventive to do that. Fortunately, Peter Sorokin was.

L&A: And what developments have pleased you most during the past 25 years?

Townes: I'm very pleased, as well as surprised, by how useful the laser has turned out to be in the medical field. To me it's very rewarding to know that my friends' eyes have been saved by lasers. It's very different from the physics I do, more personal and emotional.

The laser is a natural for eye surgery, but it's also beginning to be successful in a number of other areas. That surprised me. I wrote a paper in the early days about possible medical and biological uses, and in that paper I was not optimistic about widespread medical applications. But doctors seem to be finding many more applications for it than I had supposed. As you know, they're investigating using lasers and fiberoptics to remove plaque from the circulatory system. And there are special ways of attacking cancer with lasers, for example by producing singlet oxygen to destroy tumors.

Communication by fiberoptics is another area whose growth has been very pleasing to me, partly because that was something that was in our original patent. When Schawlow and I had a patent written up on the laser—we called it an optical maser then—we emphasized the possibility of communication partly because it was a Bell Labs patent. Now it's really coming into enormous use; it should make a very big difference in the cost of communication. Transoceanic fiberoptic cables seem to be cheaper than satellite communication. And, of course, they are much more secure.

L&A: Have there been any laser applications or developments during the past 25 years that have displeased you?

Townes: I guess one thing I don't like is the popular press harping on the laser as a death ray. Of course, death rays are fascinating to the human race; it's something they've speculated about long before the laser came along. Consider Jupiter's lightening bolt and Buck Rogers' ray gun — it's somehow fascinating to the public. But it's a distortion of the laser's overall potential for the newspapers to keep talking about lasers as death rays, when in almost all cases they are very poor weapons. I guess I find any kind of distortion displeasing, but that one is becoming pretty boring. I wish the press would do better.

ARTHUR L. SCHAWLOW

Origins of the Laser

After receiving a doctorate in physics from the University of Toronto in 1949, Arthur L. Schawlow spent two years as a postdoctoral researcher at Columbia University under Charles H. Townes before joining the technical staff at Bell Telephone Laboratories. He continued to collaborate with Townes while at Bell Labs, coauthoring a book on microwave spectroscopy and developing the principles of laser operation. In 1961 he left Bell Labs to become a professor of physics at Stanford University. Over the years his interests have evolved from developing new types of lasers to using lasers for high-resolution spectroscopy.

In 1981 Schawlow shared the Nobel Prize in Physics for his contributions to laser spectroscopy. He was elected one of six honorary members of the Optical Society of America, made president of the Optical Society of America in 1975 and of the American Physical Society in 1981, named a fellow of the American Academy of Arts and Sciences, and elected to the National Academy of Science. The OSA awarded him the Frederick Ives Medal in 1976 "in recognition of his pioneering role in the invention of the laser, his continuing originality in the refinement of coherent optical sources, his productive vision in the application of optics to science and technology, his distinguished service to optics education and to the optics community, and his innovative contributions to the public understanding of optical sciences."

Schawlow received the third Marconi International Fellowship in 1977 and the Golden Plate Award from the American Academy of Achievement in 1983. In 1982 the Laser Institute of America presented him the first Arthur L. Schawlow medal for laser applications.

C. Breck Hitz conducted this interview in early 1985 at Stanford University.

L&A: Did your interest in science and optics begin early in life?
Schawlow: I would say so. As a boy, I was always interested in technical things, particularly radio, which was the term at that time for what we would now call electronics. It's hard for a person nowadays to imagine the excitement about radio in the 1920s, when I was a little boy, and into the early '30s. When I was about five years old, we got our first radio set, which had a big old horn loudspeaker. It needed an armful of batteries to run it. I remember particularly one year just before Christmas, when the local department store had a series of broadcasts of Santa Claus' progress from the North Pole to their store. All the kids on the block would come around to listen to this on our new radio.

I remember that at least once a week the newspapers used to have a column, a technical column, on radio: How to build your own one- or two-tube set, or crystal set, that sort of thing. It was something like the excitement over microcomputers during the last few years.
L&A: That influenced you in your career choice?
Schawlow: Well, I was interested in that and also in mechanical things. I had a Meccano set, and I used to go to the library and bring home several books and read them. That's all there was to do in the summer. I went through nearly everything on the engineering and technical side of the public library.

I wanted to be a radio engineer, but there were two problems. When I finished high school, I was only sixteen years old, and you couldn't get into engineering unless you were seventeen. And there was a more serious problem. This was 1937 and the middle of the depression. I didn't have the money to go to university unless I got a scholarship. But there weren't any scholarships in engineering. I didn't quite know what to do, whether to take another year and try to earn some money, or what. Just for practice, I took the examinations for scholarships and was somewhat surprised when I got one in mathematics and physics.

I decided, "Well, that's near enough to engineering." So I went into mathematics and physics.

What I really expected to do, as practically everybody in our class did, was to teach high school physics. We had an honors course system

Arthur L. Schawlow

at the University of Toronto, whereby you chose a major right away; the course was pretty narrowly focused from there on. Your first year was practically all mathematics and physics, and then later you'd branch out into specializing in mathematics or applied mathematics or astronomy or chemistry or physics. I moved into a physics specialty.

I think all of us in that class had high school teaching as a goal. But by the time we graduated Canada was at war, and none of us actually went into high school teaching.

L&A: People wound up in the armed services instead?

Schawlow: That's right. Some of them served in the war as radar officers, or worked in war factories, or, as I did, taught courses for the armed services at the university during the war. Then I worked for a year on microwave antennas at a radar factory.

L&A: Were you planning on graduate school then?

Schawlow: You know, that wasn't automatic the way it is for a lot of people nowadays. My family didn't have a lot of money. My father was an insurance agent, and the people that lived around us were clerks and bus drivers, that sort of thing. Not laborers, but still fairly ordinary working people. Most of them had never gone to university, much less graduate school. In fact, very few people in my high school class went to university, maybe two or three out of fifty or so.

So the thought of going on for graduate work and a PhD or research career was just really too much. I hoped I could get into some radio work, somehow, but I didn't really see it very clearly. The fall-back position would be high school teaching.

L&A: How much time elapsed, then, between the time you graduated from college and the time you started graduate school?

Schawlow: I graduated in 1941, in February. Then I started back to graduate school in 1945. But I did get a masters degree earlier. I did a little research and got a masters degree in 1942. Then I didn't get my PhD until 1949.

L&A: Then what did you do, after you finished your PhD?

Schawlow: I wanted to do basic physics by that time. About a year or so before I got my PhD, there was a meeting in Ottawa of an organization called the Canadian Association of Physicists. I.I. Rabi from Columbia University gave an invited talk, and it was marvelous because he described the wonderful discoveries that Willis Lamb and Polycarp Kusch had made recently, for which they later got Nobel Prizes. I really wanted to go to Columbia more than anything else in the world. So I wrote to Rabi and he wrote back suggesting that I apply for the Carbide and Carbon Chemicals Corporation fellowship for the application of microwave spectroscopy to organic chemistry.

Well, frankly, my interest in organic chemistry was zero. I knew nothing about it. I'd never even had a course in it. But I was interested in microwaves. Actually, we'd had a klystron at the student lab in Toronto, and I'd had a chance to play with it. Besides, I had done microwave work for a year during the war.

The fellowship involved working with some guy whom I'd never heard of named Charles Townes. But I wanted to go to Columbia, so I applied for it anyway. And I won it.

Well, working with Townes at Columbia was a marvelous thing. There were no less than eight future Nobel Prize winners at Columbia then, including Hideki Yokawa, who got his Nobel a few months after I arrived. Aage Bohr, that's Nils Bohr's son, was there, as were Townes and Polycarp Kusch and Willis Lamb and James Rainwater and Val Fitch. It was a very exciting compared to Toronto, which was somewhat of a backwater then. Toronto's much better now, but it had been badly damaged by the depression and hadn't really restored itself during the first few years after the war.

L&A: How long did you stay at Columbia?

Schawlow: I left in 1951, after two years there. I started a book on microwave spectroscopy with Charlie Townes, but we didn't really finish it until 1954. Even after I'd left Columbia, I'd drive in nearly every Saturday and work on the book.

L&A: What else were you doing in the meantime?

Schawlow: Well, I was getting married and I needed to have a job. Although there had been lots of academic jobs in 1949, there weren't so many in 1951. My wife wanted to stay near New York where she was studying singing.

So when I was offered a job at Bell Labs, it seemed like a good arrangement. John Bardeen had gotten interested in superconductivity and he wanted someone to help with his experiments. Well, I had absolutely no background in low temperatures or superconductivity, but there weren't too many PhDs around then and they offered me the position.

Unfortunately, by the time I got there Bardeen had decided to go to the University of Illinois. So there I was all alone, trying to do something on superconductivity. I had to learn all about solidstate and superconductivity by myself. But I taught solidstate physics courses for their incoming engineers, and that gave me a chance to learn some of the stuff I should have learned before.

L&A: Were you involved in any of the early maser work going on about that time?

Schawlow: No, I didn't work on masers at all. I kept an eye on it, though, and I knew it was going on. In fact I had witnessed Charlie's notebook before I left Columbia. He told me about the maser idea, and it sounded good to me. If I had stayed at Columbia, I probably would have worked on it with him. But it was his idea.

By the summer of 1957 I was beginning to wonder seriously if there was some way you could get much shorter wavelengths out of some kind of maser. I wasn't really thinking about visible light yet, but of far infrared.

Charlie was consulting with Bell Labs by then. During lunch one day in early autumn we got to talking seriously about making some kind of an infrared or visible maser. We decided to put our heads together and see whether we could identify the problems and solve them.

He had one good idea right away. He thought that, instead of trying for the far infrared, we should go to the near infrared where we knew a lot more about the properties of materials. Spectra in the far infrared were pretty much uncharted at that time, you know. He told me about a scheme he'd worked out with thallium, but I looked at it and decided it wasn't going to work. If I'm remembering right, the problem was that the bottom level would empty more slowly than the upper one would fill.

Well, we realized the first problem was to get enough excited atoms or molecules. In the optical region, we faced the problem that the lifetime of an excited state is limited by spontaneous emission. But Charlie had the maser equation, and we just kind of worked it around various ways. We saw that the lifetime, the oscillator strength, didn't really matter. If the transition had a low oscillator strength, the excited atom would last longer but each one would give less gain. With a high oscillator strength, you get more gain per atom but the rate at which you supply them would be just the same. But the branching ratio did matter, that is, the fraction that would radiate the desired wavelength.

We looked up oscillator strengths in various tables, and of course there was a lot of data about the alkalis. I finally picked potassium for a stupid reason: I had a visible-wavelength Hilger spectrometer which I bought when I started working on superconductivity, to measure the thickness of thin films by multiple-beam interferometry. I could do things in the visible; it was the only optical equipment I had. The first and second line of the potassium spectrum happened to be in the visible. It's the only one of the alkalis that had that. If I had known how rotten potassium is in some other ways — it's very reactive and easily quenched — I'd probably have picked something else. But we worked out a calculation showing that we could get enough excited atoms in potassium.

The other question was: what about mode selection? We never seriously considered trying to make a resonator as small as a wavelength. You wouldn't have room to put any atoms in the thing. We had in mind some kind of a large, oversized resonator, but Charlie didn't think we had to worry too much about what the modes would be like. He said, "Well, probably some mode or modes will have higher Q than others and they'll sort of pick themselves out and dominate. Or maybe it'll be jumping around rapidly, among a relatively small number of modes."

But I wasn't entirely satisfied with that, and Martin Peter, a physicist at Bell Labs, kept telling us that we ought to find a way of selecting one mode. I had all sorts of schemes to do that. There is a page in my notebook here, it's dated January 28, 1958. . .

You know, I didn't put many things in my notebook. I really was very poor at that. I was at Bell Labs for ten years, and in that time I went through only one and a half notebooks. It seems like every time I wrote something down in my notebook it was wrong, and only the things I wrote down on scraps of paper were worth anything!

Anyhow, on this particular page I had written something about mode selection. I was thinking about using a diffraction grating for walls of the resonator, to give some selectivity. Of course, people did that ten years later in tunable lasers. I had a lot of schemes of that sort.

Eventually I got the idea that all we had to do was select a direction, and to do that we should throw away most of the resonator and just leave a little piece at each end. I thought, "Well, at least it will define the direction to within the angle subtended at one mirror by the other." When I mentioned it to Charlie, he said, "It'll be better than that because the waves go back and forth a number of times and you'll

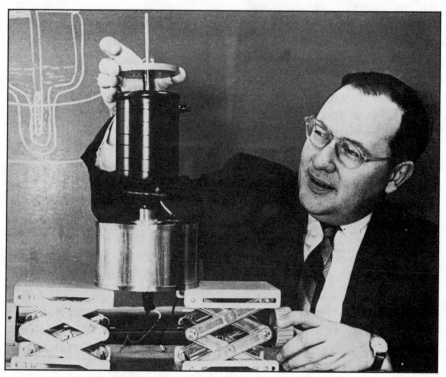

Schawlow demonstrates an early ruby laser with a straight flashlamp and cylindrical reflector, at Stanford University in 1962.

"I often joked that, no matter what you told the press about lasers, it came out as a 'death ray' or a cure for cancer!"

get higher selectivity.'' It seems very obvious when you think of it that way now — but many things are very obvious after they work.

Well, our colleagues who knew about microwave resonators didn't believe that. They wanted a calculation of some kind, which I wasn't able to do. They thought there might be some modes where the electric field was along the axis of the resonator, as there are in microwaves. Well, we couldn't answer that fully, but Charlie did some calculations with a diffraction model and showed that those modes probably wouldn't oscillate. Of course, the full theory was worked out a year or so later by Fox and Li.

You know, it sometimes annoys me a little bit when people say I rediscovered the Fabry-Perot resonator. Nothing of the sort — I'd known Fabry-Perot resonators all along because I did my thesis with one. But this was something different, because diffraction has nothing to do with a Fabry-Perot, and here it was the diffraction that gives you the mode selection. And it was really, as far as I know, the first open resonator specifically intended to select a mode.

At that point, we thought it best to publish the idea and let other people go ahead and try to build a laser. I didn't even try to build one at first. I kept on working on superconductivity, although after a little while I went to the boss and said, ''I think maybe it would be good for me to stop working on superconductivity and work on optical properties of solids.''

L&A: You were still at Bell Labs?

Schawlow: That's right. It was a wonderful place. I just had to say I thought I should stop what I was doing and start working on optical properties of solids, with a view toward developing a laser, and that was all it took.

The term ''laser,'' incidentally, is often misattributed. The name ''maser'' — an acronym for microwave amplification of stimulated emission of radiation — was invented after the first device was built. Then when masers were developed in other spectral regions, people just naturally started tossing around all sorts of variations. I've seen in print ''iraser,'' for an infrared device, ''raser'' for a radio-frequency device, and so forth. Laser was just another obvious variation. But it was the one that got popularized. Eventually we all shifted over to laser. But the important thing to realize is that it was a variation on Townes' original maser.

L&A: In the interview *Lasers & Applications* published in January

1985, Gordon Gould told us he had invented the term.

Schawlow: He was one of the first to push that name, maybe. I think I heard him use it as early as '59, but I don't think he was the first to use it. He may have forgotten how the name really came about.

L&A: While we're talking about people involved with early laser work, are there people who made significant contributions to your early work who aren't particularly remembered?

Schawlow: A lot of people were involved in the early work, and I don't think I can name them all just off the top of my head. Ali Javan thought of the idea of the gas discharge laser, and that's certainly turned out to be a pretty good idea. Bill Boyle at Bell mentioned that you could use a gas discharge but it would be better to use a semiconductor, which is a solidstate analog of a gas discharge. We didn't mention that in our paper because it was Boyle's idea, not ours. Unfortunately, he never published it either, although he got some kind of a patent on a solidstate laser.

But Javan, together with Bennett and Herriott, got helium-neon working. I think the gas laser owes nothing at all to Maiman's work, although I certainly don't want to run down his contribution, which was very important. But there still would have been lasers if Maiman had never lived, because Javan's HeNe laser was developed quite independently. Both of them did use our resonator structure, though: two little mirrors facing each other.

L&A: Maiman's laser — the first laser — operated on the R-lines of chrome-doped ruby. You had predicted that those lines weren't suitable for laser action, hadn't you?

Schawlow: Yes, it's in print. I said that in 1959. I outsmarted myself because I wasn't being quantitative.

I had the feeling that, since lasers had never been made, it must be terribly difficult to make one. You had to give yourself every advantage. In particular, it always seemed to me that you'd have to have a four-level system. With a three-level system like ruby, you'd have to lift half of the atoms out of the ground level before you got any gain at all.

But I still had ruby in mind for a laser. This was at Bell Labs, and the feeling was "Anything you can do in a gas, you can do better in a solid." So I suggested a scheme for how ruby could work. We found out that the ground level could be split by the exchange interaction between pairs of chromium atoms if the chromium concentration were high enough. And the splitting could be as large as several hundred wavenumbers. That meant that we could operate at very cold temperatures to depopulate the higher level, and in essence have a four-level system.

I tried making such a laser in a very half-hearted way. I used a dark ruby rod — you want a darker ruby because it has more of the pairs — and a little 25-joule flashlamp. Nothing much happened, so I put it aside for the time being.

That was the reason for my off-the-cuff remark, about the R-line being unsuitable for laser action. I thought I was being clever, but, as I said, I outsmarted myself. But I guess I was the first to propose ruby. It may well be that I'd drawn Maiman's attention to ruby by mentioning, in various places, that we might be able to use this dark ruby. I understand that — unfortunately — my prediction caused Maiman some difficulty. He complained that people took it more seriously than I intended it to be.

But I didn't know Maiman at all well. I'm not even sure if I had met him or not at that time. So I certainly never said that his idea wouldn't work, because I'd never heard of his idea.

L&A: Did you ever try your dark ruby laser again?

Schawlow: In 1960, after Maiman's laser had worked, I again went to the boss and said, "You know, I really believe I could make that dark ruby laser work. Should I take the time to try it?" And he said, "You owe it to yourself." So we did, and very quickly it did work. And we published it...submitted it in late 1960 or early 1961.

L&A: Twenty-five years have gone past since the events we've been discussing. What laser applications during those years have seemed most exciting to you?

Schawlow: Well, one gets certain satisfaction from the medical applications of lasers. They're always nice to point out. One of the very first laser applications, within a year or so of Maiman's laser, was repairing detached retinas. Neither Charlie nor I had ever heard of a detached retina back in 1960. If we had been trying to develop new medical techniques, we wouldn't have been fooling around with stimulated emission from excited atoms. But the eye doctors were already using xenon arc flashlamps to produce scar tissue on the retina to prevent detachment. It was obvious to them that once you had a brighter light, it was worth trying. So they did and it worked.

In the early days I used to get frantic calls and letters from people wanting me to cure them, or their relatives, of cancer. Of course, I'm not a medical man anyway. I always tried to be careful to say that the laser wasn't a cure for cancer. I often joked that, no matter what you told the press about lasers, it always came out as a "death ray" or a cure for cancer — or both! I even had somebody write and want me to cure their dog's cancer. I passed that one on to Leon Goldman, and actually he gave a very reasonable answer to what could be done for a dog's tumor.

You know, people sometimes ask me if I'm ashamed of inventing a death ray. I don't believe we have laser death rays...although I'm not sure it would be bad if we did. But we don't, at least not to do the job that people would like done, for missile defense and things like that.

L&A: There's a lot of talk in the laser community these days about the Strategic Defense Initiative. How realistic is SDI? Is that a reasonable thing to be doing with lasers?

Schawlow: Well, I don't know. I don't do military work, and I haven't since World War II. But I do think a good defense would be very worthwhile. It would be good if they could do something to end this balance of terror, which I think is inherently unstable, especially if more countries get atomic weapons. Some of the smaller countries certainly act in a very irresponsible way. At least we and the Russians can hold each other responsible. You know who it is you're dealing with. But if we suddenly had a loose missile come in, from God knows where, some kind of defense would be desirable.

I'm on the American Physical Society's directed energy weapon study group review committee. But my only function there is to make sure that the report makes sense to a person like myself, who is not involved in the SDI program.

I think, from all I've read, though, that no matter how much money you put in now, you couldn't do it. It's questionable whether you're ever going to be able to do it satisfactorily. Some people argue that the offense can always overwhelm the defense with more cheap missiles, and I don't know that they're wrong.

L&A: What other laser applications have you found gratifying?

Schawlow: There have been a lot of scientific applications. This was one of the few things we had in mind back at the beginning. When we went into this, I understood the history of radio and how people had gone from broadcast band to shortwaves, and found they could talk around the world with them. And then very high frequencies made possible broadband broadcasting, like television and FM. Then, at even higher frequencies, microwaves were no good for broadcasting but great for radar and point-to-point communication. I was certain that, if you could make coherent radiation at shorter wavelengths, there would be uses for it.

We had a vague idea about laser chemistry. We also thought you could probably do some spectroscopy with lasers, if you could learn to tune them. But I didn't think it would be easy to tune very far. Well, the first few years you couldn't tune them very far, but in the late 1960s Peter Sorokin and J.R. Lankard — and, independently, Fritz Schafer and his colleagues — developed the dye laser. That meant we could do some laser spectroscopy, and it really opened up a whole new world for us. I found that very exciting. It gave us new levels of sensitivity and resolution. It meant you could detect single atoms and photograph single atoms, a marvelous thing.

L&A: I guess the only place where laser photochemistry is really practical so far is in nuclear fuel processing.

Schawlow: It may be practical there. I think they're still debating whether that's the most practical way to enrich uranium. [The US Secretary of Energy decided in favor of the laser isotope separation process on June 5, 1985—*Ed.*] I stopped working on photochemistry because I was afraid I might find something that would make it too easy. Obviously, the big countries need isotopes separated for atomic power and also for weaponry. But it just scares me, again, if irresponsible groups, terrorists or some irresponsible countries should get access to separated and purified uranium. So far, it's difficult to enrich uranium, and I hope it stays difficult. But I'm not going to look into it.

L&A: With all the work that's gone into laser isotope separation, you'd think that somebody would have found an easy way if there were one.

Schawlow: You would have thought that about lasers, too. It really was surprising how easy it was to make them, and also how quickly they became powerful.

L&A: What other things about lasers have surprised you during the past 25 years?

Schawlow: Well, how easy they were was certainly one surprise. We were pretty confident of what we'd written, but you're never sure that you haven't overlooked something until it actually works. So I guess I was a little surprised that it turned out to be as easy as we'd said it would be.

It also surprised me that the first one was so powerful: a thousand watts. With our background in communications and microwave spectroscopy, we thought, "Well, if you can do it at all, maybe you'll get a microwatt or something like that."

L&A: What has disappointed you about lasers during the past 25 years?

Schawlow: Some things do frustrate me. For instance, here we are, twenty-five years after the first laser, and we still don't have an efficient visible-wavelength laser. It would be so much better for metal working or for chemistry than these infrared lasers. We just don't have anything that can give us more than 20 watts or so in the visible. And those lasers are so inefficient that practically all the power goes into the device, and it soon burns itself up. We spend a lot of money in this country replacing argon laser tubes.

NICOLAAS BLOEMBERGEN
Masers and Nonlinear Optics

Born in Dordrecht, the Netherlands, in 1920, Nicolaas Bloembergen came to Harvard University after World War II to complete his doctoral thesis research on nuclear magnetic resonance. He joined the Harvard faculty in 1951 and was named Gordon McKay Professor of Applied Physics in 1957. He later became the Rumford Professor of Physics, and in 1980 he was named Gerhard Gade University Professor at Harvard, retaining his position in the division of applied sciences.

In the 1950s Bloembergen pioneered development of the three-level solidstate maser. After the laser was developed, his research interests turned to nonlinear optics. His theoretical work, which laid the groundwork for the field, was first published in the early 1960s and summarized in his 1965 book, *Nonlinear Optics*. That work on nonlinear interactions also led to development of techniques for extremely high resolution laser spectroscopy. He received the 1981 Nobel Prize in Physics "for the development of laser spectroscopy," sharing it with fellow laser researcher Arthur L. Schawlow and with Kai Siegbahn of Uppsala University in Sweden, who pioneered electron spectroscopy.

Bloembergen's other honors include the 1961 Stuart Ballantine Medal from the Franklin Institute, the National Medal of Science in 1974, the Lorentz Medal of the Royal Netherlands Academy of Sciences in 1978, the Frederic Ives Medal from the Optical Society of America in 1979, and the IEEE Medal of Honor in 1983.

Jeff Hecht conducted this interview on November 5, 1984 in Bloembergen's office at Harvard.

L&A: How did you become interested in science?

Bloembergen: In Holland I went to a Latin school, a "gymnasium," which included the equivalent of the first two years of college in America. The primary fields of study were the humanities, but we had very good science teachers with doctorates in mathematics, physics, chemistry, and biology. I became interested in physics as the most challenging subject, because it combined theory with experiment and mathematics with physical phenomena. I still feel that way.

L&A: Your education must have been disrupted by World War II. What eventually brought you to the United States?

Bloembergen: They [the German occupation regime] closed the universities in 1943, and I had to hide and just try to survive until the end of the war. Afterwards I passed my qualifying [doctoral] exams in Holland, but circumstances were so difficult that I wanted to get out. I had always planned to do my thesis research somewhere else, and after the war the United States was the only place to go.

L&A: You started out working in nuclear magnetic resonance. What took you to masers and eventually nonlinear optics?

Bloembergen: It was a very natural evolution. I wanted to branch out and learn something about microwaves. I was interested in nuclear resonance saturation, so I studied the same problems in microwave magnetic resonance. The maser came along as a very special application of saturation in magnetic resonance, where one field at a high frequency saturates the resonance and causes amplification at a lower frequency. The result is what is called an inverted or negative-temperature population. My colleagues [Edward M.] Purcell and [Robert V.] Pound here had discussed these things for NMR, but never with practical applications in mind.

L&A: How did that lead to new approaches to masers?

Bloembergen: The ideas involving population inversions in NMR were only for transient cases. For the maser, I devised the three-level pumping scheme, which is a steady-state solution so you can get gain and maintain it afterwards. [Townes's first ammonia maser was a two-level system, where the inverted population was produced by separating excited ammonia molecules from those in lower states.] Three-level pumping originally involved microwaves, but clearly the scheme is very general, and it is the basis of essentially all lasers as well. Historically, that pumping scheme came first for microwave masers. Then Townes and Schawlow saw the opportunity to use it

Nicolaas Bloembergen

at optical frequencies.

L&A: Did you use a resonant cavity when you tried to build a maser?

Bloembergen: Yes, that's sort of natural, though we also had a traveling-wave type. You really had to have two cavities, one resonant at the pump wavelength and one at the signal wavelength. We built one and it worked, but it wasn't the first one. We had been concentrating on a maser amplifier for the 21-centimeter wavelength of intergalactic hydrogen, which is the prime scientific application of the maser. Unfortunately, it turned out to be harder to get amplification at 21 cm than at the X band, and the first three-level maser was one at Bell Labs that amplified at the X band after being pumped with the K band.

L&A: Your first solidstate maser was from a chromium-doped crystal of potassium cobalt cyanide. Didn't that lead to an interesting incident?

Bloembergen: In 1958, Charles Townes and I shared the Morris Liebman Award from what was then the Institute of Radio Engineers and has since become IEEE. We both went to a dinner in New York City and brought our wives along. Mrs. Townes talked to my wife and showed off a very nice pendant with a ruby crystal set in gold on a gold chain. Her husband had had it made for her in commemoration of the maser, which I think was a very nice gesture. That night in our hotel room my wife asked, "When are you going to give me something related to your maser?" So I said, "Well, dear, my maser works with cyanide."

L&A: Did you try to build a laser?

Bloembergen: No. I was well aware that in principle you could adapt the maser scheme to higher frequencies, including the optical range. But it was too risky to try with the small group here.

One reason it took us longer [than Bell Labs] to build the three-level maser was that we didn't have the support facilities here. I think all the early lasers were developed first at industrial research organizations. There was [Theodore] Maiman at Hughes with ruby, and [Ali] Javan at Bell Labs with the helium-neon laser. The carbon-dioxide laser was pioneered by Kumar Patel at Bell Labs, and the dye laser was conceived and built by [Peter] Sorokin at IBM. The semiconductor laser was at General Electric, IBM, and Lincoln Labs [a government-funded laboratory managed by the Massachusetts Institute of Technology]. What was really needed was a big support organization which could focus different technologies on a common goal. At American universities, you plod along with small groups of specialists. Harvard only provides buildings, but you really need support staff, technicians, glass blowers, opticians, and so on. History clearly shows who developed lasers first. You could try to duplicate them afterwards, but even that is hard.

L&A: How did you get your first laser?

Bloembergen: We had to wait to buy the first model, from Trion in 1961. That's what everybody else [in the universities] did, and everybody who got one of those lasers discovered plenty of new effects. I saw that we were not going to discover any new lasers, but the least we could do was to use lasers to study optical properties at high light intensities. When Maiman realized the laser in 1960, we already were working on microwave modulation of light beams. Then I decided to do modulation of light with light, which was essentially nonlinear optics.

L&A: Did you start out expecting to see nonlinear effects?

Bloembergen: Yes, but the first to see a nonlinear effect were Peter Franken and his group, who produced the second harmonic of light. It was very exciting. I was at the meeting where he announced it, and I said, "Peter, how did you do it?" Peter was very nice and modest, and said, "Nico, if you had one of those Trion lasers and you shot it at something, you would have done the same." I am not sure if I would have thought of it, but that's what happened.

In very quick succession [Robert] Terhune, [Joseph] Giordmaine, and others discovered lots of harmonic-generation properties. We had already been working on the theory when Franken published his experimental results. We already had formulas on paper. His work stimulated me, and after six months we had long theoretical papers ready, which were distributed in early 1962 and came out in *Physical Review* in April and June. We predicted reflected harmonics and so on. Reflection, transmission, polarization, refraction, and other optical effects all have analogs in the nonlinear regime, so we verified them.

L&A: How did you come to concentrate on the theory of nonlinear optics?

Bloembergen: I always liked to do theory. Also, there are quite a few nonlinear effects in magnetic resonance, and essentially the maser pumping scheme is one of them. I was aware that a high field intensity at one frequency could modify properties at another frequency. We studied light modulation by microwave irradiation and wrote a paper on the subject, which is now mostly forgotten, but all the theoretical formalisms are very similar, so we applied them to optical frequencies.

The fields of lasers and nonlinear optics would not have developed so quickly if it hadn't been for the wide body of knowledge from microwave spectroscopy and magnetic resonance. Schawlow and Townes had experience with microwave spectroscopy of molecules and gases. I had my experience in magnetic resonance. It is part of a very logical historical evolution.

L&A: The theory of nonlinear optics seems quite complex, with formulas that can run over a page long. Is it more so than NMR theory?

Bloembergen: No, NMR has developed in great detail, too. Now the basic principles in optics are well established, and we're digging deeper and finding more detail, to dot the i's and cross the t's.

L&A: What sort of experimental work did you do?

Bloembergen: We got the lasers going and verified many of the predictions. The field got so popular you had to pick topics that graduate students could do without being scooped by much larger organizations, but we did quite a bit. We provided the first experimental verifications of the generalization of optical laws of reflection and refraction into the nonlinear regime. We also measured nonlinear susceptibility and its dispersion as a function of frequency. In many cases we were the first to study the phenomena in detail. Then we got into very high light intensities where even dielectrics break down and suffer optical damage. That has turned out to be very important for high-power lasers. When short picosecond pulses came along, we could do time-resolved spectroscopy and nonlinear studies. I found optics so interesting that I dropped the magnetic resonance stuff in the mid-1960s.

L&A: What sort of technical problems did you encounter?

Bloembergen: I chose problems of an academic nature so the technology wouldn't become too tricky. I always liked small-scale experiments, because a single person can understand the details and individual measurement problems. I knew that unless we started a big program, spent big money, and got a big staff, we couldn't adequately solve problems that presented very big technological challenges.

L&A: Did you ever wish for an industrial lab with more equipment?

Bloembergen: Yes, but I solved that by consulting and by talking to industrial scientists. Sometimes I was envious, but I always had some ideas we could pursue without big experiments.

A small organization has its advantages. The working conditions are good for small-scale experiments. Students here can talk directly

Bloembergen in 1974 at work in his Harvard lab.

to the machinist in the shop, where in a big organization they have to wait and go through channels. In the 1960s the climate for funding was very liberal. We had the Joint Services Electronics Program, which let us shift from magnetic resonance to optics without any explicit proposal. It was not until the late 1960s that I had to start writing proposals and all that.

L&A: Who has worked with you in nonlinear optics?

Bloembergen: My coauthors are all listed on the papers. The early ones were John Armstrong, now at IBM; Peter Pershan, now my colleague here at Harvard; and J. Ducuing, a Frenchman who was for a time director general of the French national research organization CNRS and now is director of research for Philips in France. There was Richard Chang, now at Yale, and Ron Chen, a professor at Berkeley who just published a book, *Principles of Nonlinear Optics*, that is the first really big improvement on my early book, which I am happy to say is not wrong, but far from complete. Two of my more recent students also have written books in the field recently, Marc Levenson [of the IBM San Jose Research Laboratory] on nonlinear spectroscopy, and John Reintjes of the Naval Research Laboratory on nonlinear processes in liquids and gases.

I have had many good students here, including some other very remarkable people, and I don't want any of them to feel short-changed. I have supervised a total of 58 PhD theses and about as many postdocs over the years. I couldn't have published as much without all those people.

L&A: Were you surprised at the success of nonlinear optics?

Bloembergen: I'm still surprised. We can transform 80% of an infrared beam into the ultraviolet. And the enormous frequency resolution made possible by nonlinear spectroscopy developed by people like Ted Hansch is fantastic.

L&A: Had you expected that resolution theoretically?

Bloembergen: I never really took time to think about it, but I probably should have. I always concentrated on properties at high power, which in some cases is just the opposite of high resolution. But it isn't always. Very recently we found a line which is only 10 kilohertz wide, caused by collision-induced coherence. Everything goes just the opposite of what you might expect. The lines get both more intense and narrower the more collisions you have. Originally we predicted these collision-induced coherent effects in one of those page-long formulas you mentioned in my 1965 book. Nobody had ever seen that term then, and some people said it probably didn't pertain to reality, so I was very gratified that we demonstrated it experimentally a few years ago.

As always, if you have a new handle, you find some new things. We have been studying these optical resonance effects in phase conjugation, and it is really amazing how fine the frequency features are. We are doing phase conjugation in sodium with four or more atmospheres of argon buffer gas, and it seems that the more buffer gas you put in, the better things get. We have to find the limits, so we are going to put in 10 or maybe 100 atmospheres of argon and see what happens. It's quite exciting; there always are surprises.

L&A: Could there be other terms in your early equations waiting to surprise you?

Bloembergen: I don't think so, but some people say that if you go to higher-order nonlinear effects, such as those quartic in the field amplitudes, you still may see surprises. My colleague Tom Mossberg here at Harvard is getting new optical echoes. It is amazing how such things let you store information optically and perhaps recall it after quite a long time. Other groups are working on that also, so optical echoes still may have surprises.

L&A: Where do you think nonlinear spectroscopy and nonlinear optics will go?

Bloembergen: We have contributed very little to nonlinear spectroscopy, but I think some very basic things still may be done there. Nonlinear optics will grow more and more into applications. Fiberoptic communications is one of the biggest ones, because clearly you have to know the nonlinear optical properties of the fibers. They use the formulas we developed and analyze it in much more detail to build the best devices.

Medicine is probably the second most important application. Lasers can be used in a very sophisticated way in surgery. They even use short pulses for certain problems in the eye, to get optical breakdown at the lens without damaging the retina.

Scientifically, applications will be in ultrashort pulse research; lots of phenomena have to be investigated in the femtosecond regime. There is an awful lot of photochemistry to be done. Basic physics may be last in importance except in high-resolution spectroscopy to check gravitational theory, and to measure some more decimal places in the interaction of hydrogen atoms to challenge the theorists.

Metalworking and materials working clearly will grow substantially in the next decade. Every year the Materials Research Society meets here in Boston, and there is always a big laser component. Lasers will be used in geophysics to look for earthquake precursors and used in clocks to test special relativity.

L&A: Aren't some of those areas more general laser applications than specific applications of nonlinear effects?

Bloembergen: They are laser applications, but to get high resolutions you always need nonlinear effects. To eliminate Doppler broadening you need either two-photon absorption or saturation, which is really a nonlinear effect. It's quite surprising how nonlinearities crop up everywhere.

I feel not surprised, but very pleased, that the wide spectrum of laser applications penetrate so many different fields of human endeavor. Now there is the idea of using high-energy lasers as defensive weapons.

L&A: I was surprised to see you getting involved in that controversy, as co-chair with Kumar Patel of the American Physical Society's study group on "Star Wars," or the Strategic Defense Initiative. Did they have to twist your arm?

Bloembergen: Yes, I guess they twisted my arm. I don't relish it, but I think it is a job that has to be done. In most studies so far, most of the attention is paid to politics. They say, "Suppose we had an extremely powerful laser up there, what would happen?" Our study will be different, restricted to assessing the proposed technology.

L&A: If you had to do it all over again, would you still work in this field?

Bloembergen: Yes, it has been a good field, a very fruitful field. But my answer would be different if I were to start out as a young graduate student now. I wouldn't go into it because the field has matured. I probably would go into biophysics and biochemistry, because there you still can do some relatively small-scale experiments. If you are clever in your experimental techniques and understand the field, you can do really clever things quickly in a small group.

GORDON GOULD

Another View
of Laser Development

Gordon Gould's patent claims have made him one of the most controversial figures in the laser world. As a graduate student at Columbia University in his mid-30s, he was exposed to early maser work in the 1950s. Drawing on his earlier background in optics, he outlined a plan for trying to build a laser. Instead of trying to publish his results in a scientific journal, he chose to seek a patent. His patent applications went through many years of delays and many challenges, but he finally received two US patents, one in 1977 on optically pumped laser amplifiers, and one in 1979 on a broad range of laser applications. The validity of those patents remains in litigation.

In 1973, Gould helped found Optelecom Inc. in Gaithersburg, Md., a maker of fiberoptic equipment and systems where he became vice president and chief scientist. He retired in 1985 but remains a consultant with the company.

Jeff Hecht conducted this interview on September 19, 1984 at Optelecom.

L&A: How did you get involved in science and technology?
Gould: I always knew that I wanted to be an inventor, even before high school. I followed a course which led me in that direction, taking physics in college, then actually trying to invent. After a while, in the late 1940s, I realized that I didn't know enough yet to do the kind of things I wanted to do. For example, I had an artificial diamond project that produced a lot of graphite, but no diamonds, because I was too ignorant of the thermodynamics involved. That was when I decided I had to go back to graduate school, and did so at Columbia University.

L&A: How did you get interested in optics?

Gould: It was at Union College, where I had a professor, Frank Studer, whose undergraduate optics course just fascinated me. After graduating in 1941, I worked a summer at Western Electric, which convinced me I did not want to climb all the rungs in that big an organization. I went on to graduate school at Yale, which is very well known for spectroscopy and optics. Then came World War II, and in 1943 I asked my professor, W.W. Watson, what he might suggest in the way of essential work to avoid the draft. He told me to go to a certain address in Manhattan and tell them he had sent me. I did that the day after my draft physical, and it turned out to be the Manhattan Project. Later, when I went to Columbia, my PhD thesis on optical pumping of a thallium atomic beam got me involved with optics again, and before I finished that project I had thought of the laser.

L&A: How did Charles H. Townes's maser work at Columbia influence you?

Gould: It was very exciting, and obviously it got me to thinking along those lines, although my research project was not on a maser. I did think of an optically pumped laser, which incidentally I wrote down in a notebook and Townes witnessed. About a year later, I realized how to make a laser: a Fabry-Perot resonator would solve the problems of needing a low-loss cavity resonator for the laser light while at the same time allowing pump light to shine in to excite the medium optically. It didn't have to completely enclose the resonant light modes. I got that electrifying idea in November, 1957. About two months later, I got to thinking about this thing and knew that it was going to be the most important thing I ever got involved with in my life. I realized I would have to leave Columbia to work on it, because Professor [Polykarp] Kusch, my thesis advisor, would never let me substitute a thing like that for a research project of the very pure and basic type that was characteristic of Columbia, although Townes dabbled in such things.

L&A: A theoretical description of the laser was not pure enough?

Gould: No, there should be no practical applications. That makes it pure right off. That attitude does not exist so much today, but in those days there was a very sharp distinction between basic physics and applied physics. Columbia did not deal with applied physics.

L&A: Where had you picked up the idea of optical pumping?

Gould: [Prof. I.I.] Rabi came back from a conference in France, where

Gordon Gould

the idea was introduced, very excited by the possibility of using optical pumping along with molecular or atomic beams — which were the big thing at Columbia — to make some new kinds of measurements. I had been working on the thallium beam for three or four years already, without much success, trying to excite the atoms thermally or by electric discharge. But I never got enough into the metastable state that I was trying to measure. Rabi said to try optical pumping, and for a graduate student that was an order. I tried it and it worked, and I was able to get something like 5% of the atoms into that metastable state. That introduced the idea into my mind, and it fermented there. First I came up with the optically pumped maser and then the laser. But it wasn't so much optical pumping that was the exciting thing, it was the concept of the resonator. Optical pumping of a gas was just the first of several different kinds of excitation I thought of for laser media.

L&A: How did you get the resonator idea?

Gould: While I was at Yale, I used Fabry-Perot resonators and became familiar with the tools of optical spectroscopy. Years later I went to Columbia, which was big on microwave spectroscopy. To think of the Fabry-Perot as a resonator for a laser oscillator I had to have both those kinds of experience. It just clicked that one exciting night, about one in the morning, and I jumped up and started writing, and wrote that whole first notebook in one weekend. Then I had it notarized on Monday.

L&A: Was that all night, all weekend?

Gould: Yes, which I could do in those days. Now in the case of Townes and [Arthur L.] Schawlow, who thought of this independently several months later, the combination of experiences was in two different people. Schawlow did his thesis in optics at the University of Toronto; Townes was the father of microwave spectroscopy and the maser. People ask, "Did it really hit you like a bombshell right out of the clear blue?" Well, in a certain sense it did. It also involved 20 years of stuffing necessary bricks and mortar into my mind to a purpose I didn't know.

L&A: Once you had outlined your ideas for a laser, you tried to build one. What were the problems that kept you from having the first working laser?

Gould: The first one was money. Realizing that I couldn't do the work at Columbia, I left in 1958 to work at Technical Research Group Inc.,

TRG, out on Long Island. I told Lawrence Goldmuntz, the president, that I had some ideas which I would like to reserve out of the usual patent agreement one signs when one comes into a company. He said okay, write them out and we'll exclude them. I hadn't written up the laser very well, and I had to work on other things as well. By the time I was done it was clear that some of the work had been done while I was with TRG. Eventually we agreed on a split of rights; I would retain certain rights and TRG would also have some. Goldmuntz became excited about the project, so TRG began looking for research support.

L&A: Where did you finally get your money?

Gould: From DARPA [the Defense Advanced Research Projects Agency].

L&A: We have heard that you asked for $300,000 and they gave you a million. That's something that normally does not happen.

Gould: That's for sure. But they were exceedingly happy about the prospect of a death ray, Buck Rogers style, although I wasn't so hot on that idea. There were plenty of real applications for the laser, but that's what got them in. There was a second problem that was still not a technical problem: the project became classified and I couldn't work on it after having gone to all that trouble. I was considered a security risk because in the late 1940s I had been in a Marxist study group, so I could not get a clearance. The technical problems in building a laser were several, but I was not really working on the project, so I couldn't deal with those.

L&A: How did you interact with the people working on the classified project at TRG?

Gould: I was assigned other tasks, including several other projects which were not on the subject of lasers at all. As far as the laser project was concerned, I was there as a consultant. If anyone wanted to consult with me, they could, without telling me what they were doing. I had a fair idea of what Steve Jacobs and Paul Rabinowitz were doing with an optically pumped gas laser. But the real technical problems came because people working on the program didn't really follow my proposal at all, but set out to do other things instead of making lasers. For example, they started a great big project to learn how to grow crystals instead of taking a natural crystal and simply making a laser from it. The real reason why the first lasers were not built at TRG was simply because they didn't try.

L&A: Eventually they did build one of the first lasers, didn't they?

Gould: Yes, the optically pumped cesium laser, pumped by helium emission. The only one built anywhere was at TRG.

L&A: Whatever happened to the cesium laser?

Gould: It didn't put out very much power — about one-tenth the power of helium-neon — and the wavelengths were long, 3 and 7 micrometers. It also entailed massive problems with cesium, an alkali metal of the worst sort, so it would have been expensive to make. HeNe was much more practical, so nobody really gave it much thought.

L&A: Were you at all surprised when the laser worked?

Gould: Not at all; I knew it was going to work.

L&A: Were you surprised that Ted Maiman was the one who came up with it?

Gould: I was surprised at that; I didn't even know he was working on it. He sort of startled the world. Schawlow at Bell Labs and Irwin Weider at Westinghouse had made some measurements that seemed to show that ruby fluorescence was very inefficient, but they failed to account for self-absorption, which was happening with a vengence in ruby. I believe Maiman ground up some ruby, so the distance light had to travel to the surface was very small; it got more than 95% quantum efficiency, so he knew he had enough fluorescence to make that laser work. A group at TRG had started to work on ruby, which I had suggested among other things. Then they listened to a talk by Schawlow, who turned everybody off on ruby, and they put it on the shelf. After Maiman showed that it did work, they made one within one month at TRG. Similarly at Bell Labs they had one working within a month. They both heard about it through the grapevine before it was published.

L&A: Your patent application spelled out a number of possibilities for laser action. Was TRG trying them all?

Gould: No, even the million dollars was not enough for that, but they did select about six things. One was the potassium-emission-pumping-potassium scheme that Townes suggested, which has never worked. For seven or eight months they ignored cesium, and it was only a great effort on my part that got the projects shifted around. That might have been the first gas laser if they had started working on it right at the beginning. They also worked on a sodium-mercury discharge laser, which I had spent a lot more time discussing in my proposal than HeNe. They never succeeded in building it. For years,

Gould and colleague in their laboratory.

*"I realized it
was going to be
the most important
thing in my life."*

people opposing my patent application on the discharge laser cited the fact that nobody had built a sodium laser. Recently we decided to scotch that line of reasoning by building one. It is sitting right out there in the lab now, and it does work, just like I said it would. The people involved in my efforts to get a patent on the discharge-pumped laser came up to see it and were astonished.

L&A: Did you develop any other types of lasers?

Gould: The copper-vapor laser, in work that started at TRG which I took to the Brooklyn Polytechnic Institute when I moved there in 1967, as an instantly tenured full professor. It is an interesting and unique type of laser, which after many years of trying to solve technological problems is now on the market. Eventually it will be more important for high-power applications than either ruby or YAG lasers. Copper vapor is about 2% efficient, so to get a kilowatt you need a power supply that fills up a room. However, it is hard to make into a product because you have to heat it to 1,700° to vaporize the metal, and that takes a pretty special type of furnace and costs a lot of current. What makes it possible now is that someone designed it so the energy used to excite the discharge also heats it to the right temperature.

L&A: Do you have a patent on copper vapor?

Gould: Yes, but the market is too small to bother with now. A market has to be $20 to $30 million a year before it's worth trying to force people to license a patent, because that's going to cost you half a million dollars, even if your patent is iron-clad.

L&A: Whatever happened to TRG and its interest in your patents?

Gould: In 1965, TRG merged with Control Data Corp. and was put into an aerospace division. It lost money from the very day it merged into CDC though it had always made a profit before. In 1970 they liquidated it, but because of a clause in my patent agreement, they couldn't just turn around and sell the patents to someone else. Hadron bought the laser business but didn't want to spend the money needed to defend the patent claims, so the Control Data vice president handling the liquidation asked me to make an offer. I didn't have much money, but my lawyer suggested I offer them a dollar and 10% of whatever I made from the patents, and I got them. That was the first time I really owned them, and boy did they start burning holes in my pocket. Within three years I was out of money and trying to sell them. At that point I met up with Refac, who brought in lawyers who had been very successful before, and they started producing results, getting the first two patents [on optical pumping and laser applications] issued.

L&A: Although your patent claims have been disputed, aren't you generally credited with coining the word "laser" in your notebooks?

Gould: Yes, and there is another story I like to tell about that. I was one of the early presidents of LIA, back when the initials stood for the Laser Industry Association. The Electronics Industries Association, EIA, didn't like that because it sounded too much like their name, and because LIA was trying to work with EIA's laser group, we tried to find a different name to accommodate them. We were thrashing around the name question at a board meeting out in Arizona when Arthur Schawlow looked up with a twinkle in his eyes and said, "I know what to call it: the Optical Maser Association." Then Peter Franken said, "You lost that battle long ago." [Townes and Schawlow had used the term "optical maser" in their early papers.]

L&A: What is the current status of the optical pumping and laser applications patents?

Gould: They are being re-examined by the Patent Office at the request of General Motors, AT&T, and Control Laser Corp. We expect every claim to be rejected, based on Patent Office actions. And we also realize that something is going on in the Patent Office and that these patents are not being dealt with in any objective way whatsoever. We expect that we will appeal to the courts and that the patents will be reinstated.

L&A: How did you come to be vice-president and chief scientist at Optelecom?

Gould: By 1973 the pressures on professors like me at a place like Brooklyn Poly were getting very bad. More students, more classes to teach, and at the same time more demands for research contracts to support the graduate students. Bill Culver was trying to get me to join him to start Optelecom. Meanwhile, Brooklyn Poly had merged with the engineering department of New York University, which was losing money hand over fist because the number of students had been declining for several years and there were too many tenured professors. So the management proposed to give anybody who would resign his tenure a whole year's terminal leave with pay. Since I was going to quit anyway, I jumped at that. It was that year, which I worked without pay, that was my investment in Optelecom.

L&A: What technology are you developing at Optelecom?

Gould: I am far away from lasers themselves, except for building the little sodium-mercury laser back in the lab. At this point, I'm in an optical-communications company. I picked out a special niche so we didn't have to try to compete with the big telephone companies. That niche is developing specialized cable for oil-well logging, where you send an instrument package down a well hole and measure what is there. That can be 30,000 feet down, where the temperature is 500° F. Traditionally they have used an armored electrical logging cable, which is limited to 70,000 bits per second. That sounds like a lot, but it's not enough to carry the data that these instruments can generate, so there is pressure to get an optical fiber into that cable. We designed a new type of fiberoptic cable, and have delivered a complete telecommunication system based on it to Chevron, and they are using it. It may seem far afield from lasers, but actually we use a laser transmitter in our systems.

L&A: Is that because of the distance or the data rate?

Gould: The conditions at the bottom of the hole are too extreme for a laser. So we have the laser at the top of the hole, send the light down, modulate it down there, and send it back up. We use a Nd:YAG laser that puts out a watt or so. By the time we get to the bottom of the hole, it's down to a milliwatt, and by the time we get back up again, it's down to a microwatt, but that's still plenty.

L&A: Does the present state of laser technology surprise you?

Gould: The only laser that did surprise me was the semiconductor laser; it surprised me that it was feasible. Other laser developments, even though I may not have thought of them, didn't surprise me; they seemed rather natural when they came along, like the copper laser. I particularly am not surprised by the use of really high-powered lasers for industrial materials processing because I envisioned doing that right from the beginning in my patent application.

L&A: Where do you think it's going?

Gould: Way back in the beginning, I used to say that lasers are to light technology what vacuum tubes or transistors were to electronics. We certainly have moved in that direction, but we haven't gotten there yet. One day the field that you might call optronics may be as big as the whole electronics industry is today. The laser industry passed a billion dollars a year a while back, and it will continue to grow in volume and in the number of different kinds of applications, which will become increasingly pervasive.

L&A: Have any applications surprised you, say laser fusion?

Gould: No, because in my very first notebook I wrote: "The powers that will be available with the media, that I can see are feasible, will be able to heat an object up to 100 million degrees." So I wrote down the possibility of starting nuclear fusion. They've been working on it for 25 years and it's not here yet. But it doesn't surprise me.

L&A: What do you think about laser weapons?

Gould: You remember the reason that DARPA got so excited was because of the concept of laser weapons. What has not surprised me is the fact that they don't exist yet, or they're not operational yet. My reaction to the death-ray idea was that light is not really suitable because its wavelengths are too long. Diffraction spreads light waves out too fast for them to be effective at any great distance, and really you should be using much shorter wavelengths so the diffraction is negligible. In fact, how about matter waves? How about a bullet? That was my reaction way back in the beginning, and it is still my reaction. I don't think that much of the "Star Wars" concept. It is particularly dangerous to get people's expectations up beyond what may be feasible. They'll think that the security of the country is being improved when it really isn't. It may even drive the Russians to do things neither they nor we would want them to do.

L&A: If you had it all to do over again, would you still work on lasers?

Gould: I certainly would, but I would do it differently than I did before.

L&A: Would you look for a different lawyer than the one to whom you first showed your notebooks, and asked how you should go about getting a patent?

Gould: That would be the first thing. Just think, if I had understood him and if he and I had communicated properly in January 1958, this whole history would have been entirely different. I would have had my patent long, long ago, and it would have run out long, long ago. I would have made, maybe, $100,000, much less than the patent has brought me now. But certainly the laser proved to be what I realized it was going to be, the most important thing I ever got involved with in my life, and if it happened to me again, I would do it. Of course, at that moment in my life I was too ignorant in business law to be able to do it right, and if I did it over again probably the same damn thing would happen.

THEODORE H. MAIMAN

The First Laser

Ted Maiman's career has always been marked by more than a bit of iconoclasm. While theoretical physicists wrote papers and debated the merits of different materials and approaches for a laser, Maiman set out to build one. A pragmatic scientist, one of his main concerns was practicality — he not only wanted to build a laser, but he wanted the device to be easy to work with. Along the way, Maiman had to overcome obstacles put in place by scientific colleagues, supervisors — and the laws of physics. He was determined, however, to let nothing stand in his way. It was this independent attitude that helped Maiman win the race to build the first laser. In May, 1960, Maiman demonstrated laser action from a ruby crystal while working at Hughes Research Laboratories in Malibu, Calif.

Maiman left Hughes in 1962 to found Korad Corp., one of the first manufacturers of laser equipment. From 1976 to 1983 he was vice-president of advanced technology for TRW Inc.'s electronics and defense sector. He is currently a consultant in Marina Del Rey, Calif., and a director of PlessCor Optronics. Maiman received graduate degrees in both engineering and physics from Stanford University, where he studied under Nobel Prize winner Willis Lamb.

He is a fellow of both the American Physical Society and the Optical Society of America, a member of both the National Academy of Sciences and the National Academy of Engineers, and a recent inductee to the National Inventor's Hall of Fame.

Jim Cavuoto conducted this interview on November 14, 1984, in Marina Del Rey, Calif.

L&A: How did you first get involved in laser research?

Maiman: I did some microwave optical experiments during my PhD work at Stanford, where I was looking into the fine structure of the excited states of helium. I devised a measurement technique to analyze optical structure using a combination of electronics, microwaves, and optical instrumentation. The process used a parallel plate resonator coupled to a low-pressure helium discharge.

Later, I went to work for Hughes Research Laboratories in the newly-formed Atomic Physics Department. The real charter and interest of the department was to generate higher coherent frequencies than were currently available. This was about the time that the ammonia maser came about. Hughes had an intense interest in maser research at that time. But at first, I worked on a contract I brought in which used a nonlinear aspect of cyclotron resonance to generate harmonics.

L&A: What was your contribution to the maser work at Hughes?

Maiman: I wanted to work in a research capacity, but the Army Signal Corps, which had sponsored the original basic research, phased out its support. They now wanted a practical x-band maser delivered to them. They didn't want any state of the art advances. I was asked to head that project. I wasn't too enthusiastic at first because that was a practical device and I was more research-oriented. But then I got into it. And even though they hadn't asked me to make any tremendous advances, I decided I had some ideas to make the maser more practical.

Masers at that time had two serious practical problems. The main difficulty was cryogenics — a solidstate maser needs liquid helium temperatures. The other problem was that the conventional maser used a huge magnet weighing about 5,000 pounds. Inside this magnet you would put a dewar within a dewar. The outer dewar had liquid nitrogen in it, and the inner one was filled with liquid helium. You then put a small maser crystal in a cavity down at the bottom of the dewar, between the pole faces of the magnet. I decided that there were a couple of things I might be able to do in one fell swoop.

The preferred maser material at that time was ruby. There were a number of other materials that worked, but this was a rugged hard crystal. The others were more fragile. So I made a miniature cavity out of ruby. Then I painted a highly conductive silver paint over the ruby and put a small coupling hole in it. I decided that instead of putting the double dewar inside of that monster magnet, I was going to

Theodore H. Maiman

put a small permanent magnet down inside the dewar. That wasn't supposed to work for several reasons. One was that you weren't supposed to put a permanent magnet into liquid helium. It would crack and break. Another was that you couldn't fill the cavity with active crystal. Needless to say, it worked nicely.

So our delivery — the whole thing, magnet, dewars, and everything — was 25 pounds instead of 5,000 pounds! And it was much more stable and had 10 times the gain-bandwidth factor of previous masers. I got a lot of satisfaction out of that work.

Later, I made an even smaller maser — the whole dewar and everything was under 4 pounds. I was able to make a "hot" maser — the performance was so high, I could afford to let the temperature rise and use liquid nitrogen cooling. So I had the first liquid nitrogen solid-state maser. It even worked — marginally — at dry ice temperatures.

L&A: Was your commitment to ruby as a laser material at all influenced by the success you had with it in your maser work?

Maiman: No, not at all. I didn't know what material I was going to use at first, but I was thinking that I wanted it to be something fairly rugged. After my experience with the maser, I did not want anything that had to be cooled cryogenically.

I used ruby just as a starting point. I constructed an analytical model and made some detailed calculations. I could see that it would be very difficult to make it work. Further, Weider had published a paper which indicated that the quantum efficiency of ruby was maybe 1%. If that was true, it would rule out ruby for sure. I looked at some other materials, like gadolinium-doped crystals, because they displayed some very sharp lines. In the case of gadolinium, though, the problem is that, not only is the fluorescence very sharp, but the pumping bands are very sharp. In addition, it's in the ultraviolet, and it's hard to get much pumping energy. I went back to ruby, not to try to work on it as a laser, but again as a model, because I liked some of the other properties.

L&A: Were you suspicious of the calculation on the quantum efficiency of ruby at this point?

Maiman: No I wasn't. In fact, quite the contrary. I decided to use ruby because it had other properties that were interesting. It had a simple energy level scheme, was rugged, and in principle I wouldn't have to cool it below room temperature. To make it work, it wouldn't have to be very big — you didn't need a 1-meter cavity, as the discharge people were proposing. I decided to find out what was wrong with ruby, and then, working with some materials experts, use that as a clue to find another material that would have similar properties with the desirable qualities, but without the bottleneck — the low quantum efficiency. I made some detailed measurements to find out where the bottleneck was, and I devised a model to show what was going on.

As I went through the model and looked through all of the possible problem areas, I couldn't find the problem. In fact, I found the quantum efficiency was fairly high, about 75%. So, armed with that, I started to think more seriously about ruby. I had already looked at a number of other solid materials, and I became rapidly prejudiced against gases, at least for the first laser, because I felt the processes going on were too complicated. (The early gas lasers weren't well understood until some years *after* they worked.)

I started to get a little more interested in ruby when I found the quantum efficiency was near unity. I knew it would still take a lot of power, though. Also, I wasn't going to dismiss the supposed problem of having to depopulate ruby's ground state without studying it in some detail to see what was really involved. So I went through the calculations and came across the fact that for such a system what was really important was the brightness of the pumping source — which amounted to the equivalent of a black body of some 5,000 K. This was not unheard of, but it was really pushing the specifications of the common laboratory lamp.

I first analyzed a design with a straight mercury lamp and a ruby rod in an elliptical cylinder and concluded that operation might be possible, but very marginal. (Incidentally, several years later Evtuhov at Hughes did make such a cw ruby laser.)

L&A: I imagine practical problems like that played a very important role in the development of the first laser.

Maiman: Absolutely. My calculations told me the laser would work, at least marginally. But that wasn't good enough. I decided that, if I had to go to all the trouble of putting the experiment together, it would have to work. With the margin that close, I was very worried that the laser wouldn't quite lase. What if I was just below threshold? That would have been very frustrating.

My insight regarding the high lamp temperature proved to be very important. I remember reading that the effective temperature of a xenon flashlamp was 8,000 K. Also, I didn't see any reason why I had to do this continuouswave — pulsed mode was perfectly fine. People do a lot of things purposely with pulses, for example, radar. Besides, I was just trying to demonstrate that this could be done, not find the ultimate system. But I was determined that it not be cryogenic from that experience with the maser. Another consideration that kept me hanging in with ruby was that, if it worked at all, it would be small and compact and operate at room temperature. I was also intrigued by the fact that, if it worked, ruby would emit visible light. I would be able to see it. Essentially all of the other proposals were in the infrared and so were the subsequent early successful lasers, including helium-neon. I decided to think more about the xenon lamp.

I went through the catalogs of available laboratory lamps, and calculated the effective brightness rather than being concerned with the size or the total number of joules delivered, because my calculations showed that, within reason, the size of the ruby would not be important. And as I screened them and isolated them, there were only three lamps that had enough brightness to do it. All three of them were made by General Electric and all three of them were quartz helices. I bought a few versions of all three: FT503, FT506, and FT624 (a monster, designed for aerial photography). They all had enough brightness. So my first laser used the smallest of three lamps, a GE FT506 lamp. The ruby was about 1 centimeter in diameter by 2 cm long, just filling the lamp spiral.

That brings up a humorous anecdote. When the Hughes public relations people took the photographs of my first laser, they used the FT503 flashlamp because it was more photogenic. So when the press release got circulated, everybody thought that was the lamp I used. There was a run on those lamps. All of the reproductions of the ruby laser made in other labs used the FT503.

L&A: How did you go about selecting the appropriate conditions and parameters to use in your experiment?

Maiman: It's important to note that there was engineering design involved here. The parameters I needed to know weren't published, so I had to find out what they were. For example, what was the absorption coefficient in the different directions with respect to the crystal axis? Once I found that the calculation of ruby's quantum efficiency wasn't correct, I checked all the other parameters too. After that I didn't take anything for granted.

I did an experiment to check the validity of my assumptions without really trying the laser yet. I used my parallel plate resonator again, as I had previously with the maser. I made a cube of ruby and formed some plates that were larger than the crystal on either side of the crystal. Then I used my knowledge of the maser. The ground level of chrome-doped ruby is split into four levels by a magnetic field. The separation of those levels in zero field is about 11 gigahertz. I devised the cavity so that it would be resonant right around the 11 GHz ground state of ruby. I knew that once I was real close, just at a very small magnetic field, I could tune in — which is what I did. If I hit that crystal with the pumping lamp, then I was going to depopulate in the ground state and raise the Q of the cavity. I wanted to know if the amount of light I was putting in would produce the kinds of depopulation that I calculated would happen.

This experiment was not set up to get lasing going. In fact, I had the light come in through a light pipe — a quartz rod — which was near the flashlamp and coupled to the cube. I wanted that separation so that the electronics — pulse currents, etc. — would be away from the microwave instrumentation.

At any rate, I did observe the reduced absorption of the ground state, which I expected to happen because the light kicked the atoms up into the metastable state. The magnitude of depopulation was about what I had calculated. So that part checked. I was also able to check some of the other properties that I was concerned about.

L&A: How did you come to place the ruby crystal inside the flashlamp?

Maiman: The spiral lamps were the ones that had those high brightness temperatures. High brightness straight lamps weren't available at that time, so I couldn't use my elliptical cylinder design for a continuous-wave laser. Because of their odd shape the helical lamps weren't very amenable to reflectors. In fact I started to try to devise a focusing reflector to direct the light onto the ruby. I was talking to a distributor of GE lamps, and he told me the big one, the FT624, was so intense that it would set a piece of steel wool on fire.

Then all of a sudden it clicked in my head. Of course! The best I could do if I put that lamp in a reflector and then collected and refocused it back on the ruby was the brightness of the lamp itself. I thought, "Instead of remotely transferring the brightness of the lamp to the ruby, why don't I put it in near proximity?" I'd still be at the same brightness. I put a reflector around the outside of the lamp to collect the radiation that would travel outward. It acts as a radiation shield and, in principle, increases the brightness of the lamp for the same input power. There's the design.

And of course it came out extremely simple in realization. As with the maser, I painted the cavity right on the ruby — in fact silver again, only it wasn't paint. I evaporated silver onto the endfaces. Multilayers were around, but they were in early development. They didn't exactly have spectacular performance, and they were difficult to work with. I wanted something that was easier to work with. At first, I wanted to put a full silver on one end and a partial silver on the other. But then I realized that had experimental problems because the half silver or partial silver was extremely thin. Any tarnishing of silver changes its transmission coefficient properties. That's when I got the idea, again relating back to the microwave era, that since it was a cavity, I'll just put in a coupling hole. So I put full silver on each end and then bored a tiny coupling hole to observe what was going on. Later I plotted the output versus the hole size and found the optimum hole size. That was the progression.

L&A: It seems from your experience that there was a large disparity between the paper world of the scientific literature and the practical world of the laboratory. But were you at all influenced at the time by what was being published in the literature?

Maiman: Well, I saw the proposals and I thought, "Fine, go to it, have fun, because you have a lot of problems there." Looking back on it, I was fairly cocky, because I felt I had a sound basis for doing what I was doing. One of the nice things about simple fluorescent solids is there are only a few levels in there to account for. You can really keep track of what's going on. In the discharge laser there are so many variables to keep track of — the electron temperature, the gas temperature, the excitation cross sections, how many ions do you make, the velocity distribution, etc. Hundreds of different levels. And once you excite something to one level, another electron can come and knock it up into a higher state, then it cascades down. Nobody, even today with computers, can tell you in detail the exact processes that happen in any discharge. It's really tricky to do.

The Schawlow-Townes alkali vapor laser didn't work and, as far as I was concerned, it was pretty obvious: either it wouldn't work or it would be very, very tough to pull off. There were a number of things that they left out in their analyses. The theoretical concept was fine, but they didn't bring in the practical considerations. They looked at a pumping lamp of potassium. The parameters of the lamp are quite different from those of the absorption cell. The lamp is at higher pressure. At 10 amps, the gas temperature is higher and these parameters, plus Stark broadening, give rise to a much broader line in emission than the absorption cell — between 3 and 10 times as much. Their own calculation admits to being short a factor of 3, so now they're off by a factor of 10 to 30. They didn't take into account the fact that there's some kind of coupling efficiency — you don't get all of the energy out of the lamp into the cell. Altogether it looked like that proposal was between 1 and 2 orders of magnitude short.

But that was the kind of practical limitation they faced, to say nothing of the problems of dealing with alkali vapor in the laboratory. It's a very reactive material and difficult to handle. The same arguments apply to Gordon Gould, who proposed the same basic concept with sodium and also didn't take into account the practical parameters.

One of the things that was really important is taken for granted now. Keep in mind that nobody had ever made coherent light. One prominent physicist asserted that you could *not* make coherent light. We had pushed up the coherent electromagnetic spectrum by the late 1950s to 30 GHz or so. Generally speaking that was the practical high end. Now we were contemplating jumping by four orders of magnitude!

I became intent on finding a system in which I could get all the practical problems out of the way first, and thereby confront any unanticipated, inherent obstacles to making a laser. If I was going to fail, I didn't want it to be because of mundane factors like corrosion from alkali vapor or not having enough pump power.

L&A: Did you find yourself getting caught up in a race with other research teams to build the first laser?

Maiman: I would say that I made the commitment around August, 1959, to work on the laser. I knew there was a lot of interest. I knew what people were doing, but I felt they were going off in tough directions. Schawlow and Townes must have started their work in the spring of 1958. Here it was August, 1959, and I was aware of the fact that people all over the world were now in this race. It was a little brash for me to enter that race at that time. People with well-funded efforts had already been going for, let's say, a year. It's interesting to note that Hughes' total expenditures in our nine-month laser effort amounted to about $50,000. Contrast that with the $500,000 to $1 million that

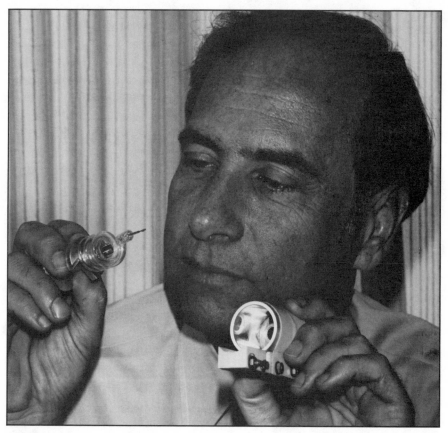

Maiman examines the first ruby laser, built at Hughes Research Laboratories (circa 1960).

"Keep in mind that nobody had ever made coherent light. One prominent physicist asserted that you could not make coherent light."

other research teams were spending.

L&A: Did this competitive atmosphere create an intense sense of pressure for you?

Maiman: Well, when I attended the first IQEC meeting, there was talk about lasers, gas discharge proposals, in addition to the alkali vapor. Schawlow gave a paper in which he made his famous comment on how ruby couldn't work because the ground state would have to be depopulated. Another one of his comments was, "Nobody knows what form the laser will take." What I mean is, some frustration had already set in. People had many proposals and had already tried some of these things and they weren't working. The alkali vapor system by now had been worked on for at least a year at Columbia University. TRG had a $1 million contract out of DARPA and worked on several different systems in parallel, all kinds of proposals that Gould had suggested. At this time, those efforts were frustrating. IBM had some ideas — I think they were working on solids even before I built my ruby laser.

Charles Townes' comments that it turned out to be easy to make the first laser and that anything will lase if you hit it hard enough are incredulous statements to me. If it was so easy, why didn't Columbia, Bell Labs, or TRG pull it off? They each had a head start, plenty of money, and heavy staffing.

Of course there were also people at that time saying it was impossible to make a laser. I had one scientist visit me who was planning to teach a course at the University of Michigan on why the laser could not be made. Every once in a while, you'd hear rumors that somebody had built a laser or was very close. They all proved to be false. But there was a fantastic amount of excitement in just the thought of winning in a race that had this quality of participants.

L&A: Did your colleagues at Hughes fully understand the implications of what you were doing? Did they lend you the support you needed to do your research?

Maiman: There was not a great deal of enthusiasm or support. Of course, all hell broke loose after we got the laser working, but that's another story.

I don't want to fault Hughes too much. In a lot of ways, this was just a symbol of the attitude you often find in a big company. There is a tremendous resistance to anything new and different.

People at Hughes questioned the money. Was it worth the company's investment to do this work? Also, the government was becoming more stringent with R&D funding and demanding tougher and more focused R&D. I was working on general research funds (known today as Independent Research & Development). It comes out of an allowable expense on big contracts, but there wasn't that much money around.

There were several criticisms thrown at me. "Who knows if you can make a laser?" "Who knows what form it is going to take?" "There are all these other people who are ahead of you who know what they're doing." "We have limited funds and we don't know what you would do with it anyway if you got it." "There's this new field of computers — why don't you work on that." "Besides, Schawlow said that ruby can't work, so the specific approach you are working on is wrong."

It became an uphill battle. I got pretty stubborn. I had seen some of the things people had done on general research funds. I felt that, if anyone had earned the right to it, I had. I had done really well on all the government contract work and brought in some contracts for them. I felt I was entitled to use of the funds. I was just carried away with that project. I knew there was something there, and I was determined to follow through on it. I had a sixth sense that it was going to work. The more I worked on it, the more convinced I became. I understood the reasons why it wasn't supposed to work. I wasn't daunted by them, though, because I felt that I knew the answers.

L&A: Do you feel that your demonstration of the ruby laser hastened considerably the building of the other types of lasers?

Maiman: Once you knew you could make a laser, all of a sudden there were a lot of lasers. For example, Stevenson and Sorokin at IBM made two solidstate lasers fairly quickly thereafter. One was samarium-doped calcium fluoride and another was uranium-doped calcium fluoride. They put crystals in an FT503 flashlamp, but they had to cool them to liquid helium to make those crystals work.

Schawlow made another version of a ruby laser. If you dope ruby heavily enough, it acts like a four-level system. The only problem is that it takes as much energy to make it go as normal ruby — plus you have to cool it to liquid helium. Having four levels doesn't guarantee low pump power.

The fourth laser was Javan's infrared helium-neon laser. (The familiar red HeNe wasn't developed until two years later by Rigden and White.) It was not the second laser, as many people think. Apparently, it was not doing that well. I don't know if people started to doubt whether you could make it or the approach just wasn't working. The point is, the staffing was beefed up when Herriott and Bennett joined Javan's project, and they still had to do a lot of experimental trials before they could figure out how to make HeNe work. I suspect that a lot of that extra motivation came from knowing you could make a laser and that there was a laser in existence.

L&A: How long did it take for practical applications to emerge?

Maiman: There were a lot of doubts about how useful the laser would be after that. You see, it was slow in coming into practical use outside the laboratory. When I say outside the laboratory, you can't toss that off too quickly because there was a rash of laser use in scientific experimentation.

But the trade magazines were looking for some practical applications. How much can you talk about Raman spectroscopy and plasma diagnostics and other exotic areas? As a scientific tool what the laser provided was just tremendous. But many people questioned whether it could be used outside the lab in a practical way. I had editors of magazines interview me and say, "Yes, but where is it being used?" People would announce applications on an experimental basis, but nobody was putting it on a production line. They didn't know enough about the proper parameters to accomplish anything in a reproducible and economic manner. The laser was, in effect, a solution looking for a problem.

People didn't appreciate that the laser offered such a radically different way of doing things that applications were bound to be slow in developing. After all, it took 30 years to get from Kitty-Hawk to commercial aviation. Of course the electronics people had an easier time with the advent of transistors. For them, it was just a matter of replacing vacuum tubes in the beginning.

With one exception, that wasn't the case with the laser. The exception was repairing detached retinas. Xenon lamps were already on the market to repair detached retinas, so it was just a case of replacing them with lasers.

L&A: What problems do you have with the state of patent laws as they exist today?

Maiman: Well, one problem in my mind is that, if a potential innovator goes to work for a large corporation, he or she automatically signs off the rights to patents that may come out of their work. The company argues that everything is done on their time, with their equipment, and they paid for it. That's too easy an answer, because there are many people working for that company who don't produce patents, and they get paid similar salaries. Also, this can provide an unusually large benefit to the company in proportion to the amount of time and effort the person has put in. As a consequence there is not strong motivation to file patents in a corporate environment. There's not enough of the monetary reward to make it worthwhile.

In some other countries, for example Japan and West Germany, the inventor has to get a piece of the action. To me, that makes a lot of sense — it's going to benefit employers of inventors because more patents will be filed. If you feel that this is something that you're going to get some benefit out of, if it really works out, then there's much more motivation. So everybody gains.

One improvement in patent laws was passed at the end of the Carter administration. That gave the Patent Office a chance, under certain conditions, to reopen a patent that has already been issued. The purpose of this was to save multi-millions of dollars in litigation costs — 80% of patents that are challenged in court are thrown out. Obviously, a lot of legal fees are being generated that might be obviated. As you know, the Gould amplifier patent has been re-examined and disallowed by the Patent Office.

L&A: What laser applications might you expect to develop in the next 25 years?

Maiman: There's been a fantastic upsurge in medical applications. These to me are so exciting and impressive. Fiber-directed cauterizing of ulcers, for example. You can now pretty much cure early cases of cervical and bladder cancer. In surgery, there's little or no bleeding, less trauma, less scarring, less post-operative pain. One of the first non-laboratory applications of lasers was in repairing detached retinas. A vast array of laser uses in medicine are emerging. The possibility of a laser cleaning coronary arteries could obviate bypass surgery. Even more exciting is the work in photodynamic therapy, whereby malignant cells are selectively destroyed. I would expect that there will be at least one laser in every doctor's office by the turn of the century.

Industrial use still has a long way to go. Heavy mechanical manufacturing is in many respects a backwards industry. We've just scratched the surface there. I would expect much greater industrial use when lasers and robots are coupled together.

It's clear to me that optical storage is at the stage of going over the hump. The computer industry's impact on that is just tremendous. I would expect both archival and erasable optical memory to dominate computer and image storage. With ultraviolet lasers, we should be able to get hole spacings of 0.1 micrometers on an optical disk, which corresponds to 10^{10} bits/cm^2 or 100 gigabytes capacity. With this degree of information density, we could store a three-hour movie in very-high-definition TV on a 5-inch disk. Now, when you couple this areal capacity with three-dimensional storage, which may become possible in the next 25 years, this yields 10^{15} bits/cm^3. By comparison, the human brain has been estimated to store 10^{14} bits — which is about the size of the Library of Congress' holdings. The same 0.1μm resolution could very well apply to microlithography of integrated circuits. Looking farther down the road, I wouldn't be surprised to see fully optical computers developed within the next 25 years.

Also, laser TV could materialize soon. It would be the ultimate TV set because you would get perfect registration and extremely high resolution. Laser light provides fully saturated primaries for color presentation.

In laser technology, we could soon see semiconductor lasers — emitting at say 0.4 to 4 micrometers with perhaps 10 W of power — dominating the industry. Their small size, high efficiency, and multi-gigabit modulation give them tremendous appeal. Free electron lasers should also become a mature technology in the next 25 years.

The technology of fiberoptics is established and the market is just starting to move at a rapid pace. I would expect it to be the dominant link for computer networks. Finally, the technology of integrated optics is just beginning to blossom. I would expect to see a tremendous upsurge in integrated optical chips over the next 25 years.

PETER SOROKIN

The Second Laser and the Dye Laser

Peter P. Sorokin was born in Boston, the son of a sociology professor at Harvard University. He studied physics as an undergraduate and graduate student at Harvard, receiving his PhD in 1958 for a dissertation on nuclear magnetic resonance. In 1957 he joined the research staff at the IBM Thomas J. Watson Research Center and in 1968 was named an IBM Fellow. He continues to work in laser science at IBM Research. In 1960, he and Mirek Stevenson demonstrated the solid-state uranium laser—the first new laser to operate following Theodore H. Maiman's demonstration of the ruby laser.

In the mid-1960s, Sorokin and John Lankard studied the optical properties of organic dyes, first developing the saturable absorber Q switch, then the dye laser. More recently he has worked in nonlinear optics and spectroscopy, developing techniques including four-wave mixing and time-resolved infrared spectroscopy, and pursuing a long-standing interest in the two-photon laser concept. A fellow of the American Physical Society and the Optical Society of America, he has been elected to the American Academy of Arts & Sciences and the National Academy of Sciences.

Jeff Hecht conducted this interview on September 27, 1984, at the IBM Thomas J. Watson Research Center, Yorktown Heights, N.Y.

L&A: How did you get involved in physics?
Sorokin: I owe that to Professor [Nicolaas] Bloembergen, my thesis advisor at Harvard. I had planned to go into theoretical solidstate physics. In my second year of graduate work, another student, Don Weinberg, and I signed up for a reading course on nuclear magnetic

resonance given by Bloembergen, which we thought would be an easy course. At the time Bloembergen wasn't a very polished lecturer, and we let everything go over our heads, but we weren't too concerned. Then he announced that he wanted a term paper from each of us as proof that we took the course, and we wrote the papers and handed them in. We got unsatisfactory grades, so we went to Prof. Bloembergen, who said "These papers don't say anything about what I was teaching."

We couldn't get unsatisfactory grades, so I spent part of the summer trying to understand NMR, then wrote up a term paper which Bloembergen accepted. By that time, I felt I had invested so much time in the subject, which actually seemed interesting, that I might as well sign up and do a thesis with him. So did Don.

First Bloembergen assigned me a theoretical problem, and for a year I sat at a desk in the Gordon McKay Laboratory with a pad of paper. Finally I came back to him and said, "The divergent parts cancel, and all you have left are terms that are very hard to evaluate, but they're finite." And he looked at me and said, "Well, Peter, I think you'd better do experiments."

I was a little shaken, but I followed his advice. Then I was issued what was called a Pound box [Pound-Knight-Watkins spectrometer for NMR measurements] and told that a suitable thesis topic would be to measure the cesium chemical shifts in cesium halides. However, the cesium resonances appeared to have long relaxation times, which caused the NMR signals to saturate and disappear on resonance. It was very difficult to measure the chemical shifts with any accuracy. Another year went by, and I was beginning to get discouraged. I would come home to my parents' house late at night, often quite dispirited. My father, a professor of sociology at Harvard and an early riser, would start a tirade early in the morning, asking, "What sort of an ignoramus are you?" He was worried that I might have picked the wrong field. In his department, if a person had been in graduate school three years, the faculty met privately and said, "This person is probably hopeless. What do we do with him?" I had been in graduate school for four years, and I wasn't getting anywhere. I really thought I might quit.

At that point, a young scientist named Al Redfield came to Bloembergen's lab as a postdoc. He had just published a theory of saturation of the absorption and the dispersion of NMR, and he impressed all

Peter Sorokin

the graduate students with his ability to design electronic apparatus. He built his own version of an NMR spectrometer, based on the crossed-coil approach of Prof. Felix Bloch's group at Stanford. I began to sense that the cesium halides might be suitable for demonstrating aspects of his theory. Also, a pulsed version of his nuclear spin induction spectrometer might be very suitable for studying the easily saturated cesium resonances (via free induction decay).

I dropped my attempt to measure these resonances with the Pound box, and built a double resonance, crossed-coil apparatus. One day, right after supper, I went back to the lab, turned on my newly built equipment, and observed an enormous signal. I repeated the experiment and discovered what caused the signal enhancement. I got really excited and instantly realized that I had a thesis. I showed this to Bloembergen the next day, and he, too, got all excited. From then on, he came every day to the lab to check my progress and advise me. In about half a year, I finished my thesis, and Bloembergen and I published a paper based on it. That experience set my course in science. The fact that I almost quit, but then found something just following my own hunches, gave me a lot of confidence, and I've retained that confidence all through my career.

L&A: How did you get interested in the laser, or the "optical maser" as it was called in those days?

Sorokin: I was subsequently hired by IBM to work with Dr. William V. (Bill) Smith on microwave resonance in solids. When the famous paper on optical masers by [Charles] Townes and [Arthur L.] Schawlow appeared in the December, 1958 *Physical Review*, Smith suggested that we redirect our efforts to this new field. Along with Mirek Stevenson, who had obtained his PhD with Townes a couple of years before, and who had been hired by Smith's group at about the same time I was hired, we decided to get involved with this "optical maser" idea. That, incidentally, was Townes-Schawlow terminology. Gordon Gould hadn't coined the name "laser" yet, but he soon would. After he proposed the term "laser," Bell Labs refused to use it and in effect discouraged others from using the term, but it won out anyway.

We all went to the first quantum electronics conference [in mid-1959], held at the Shawanga Lodge in the [Catskill] mountains, right across the Hudson River. When we came back, we decided to drop immediately what we had been doing in order to focus on the possibility of finding a solidstate laser material. We were thinking of continuous optical pumping schemes, with lamp power levels on the order of watts. I was doubtful that mirrors with high enough reflectivities could actually be fabricated, so I conceived of using a polished square [of laser material].

The idea was that the light could make many passes through the medium, bouncing off the edges of the square in a low-loss manner by total internal reflection. If you then bevelled off just a tiny bit of a corner, that would give you a way to couple out light from this resonator. A low-gain system might thus be made to oscillate. Bill Smith and Ralph Landauer [also at IBM] thought it was an interesting idea, and Smith additionally pointed out that it could be made mode-selective. If the refractive index were just slightly greater than the square root of two, only those modes corresponding to waves exactly incident at 45° on the edges of the square would have low optical loss. Calcium fluoride has an index just slightly greater than the square root of two, so I said, "Aha! We'll try to find an ion we can pump in a laser scheme and see if it can be incorporated into a calcium-fluoride lattice." Rare earths looked good because of the protected $4f$ shell, so I went through all the journals to find out who had put rare earths in calcium fluoride. In the Russian literature, [P.P.] Feofilov had reported two very striking systems in calcium fluoride. One was trivalent uranium, very similar to a rare earth, with fluorescence at 2.5 micrometers. We figured that to have a low threshold, you should pump to a broad band, then drop to a metastable emitting level, then emit to a thermally unpopulated level.

L&A: The classic four-level scheme?

Sorokin: Yes. We realized from Feofilov's published fluorescence spectrum that most of the uranium emission was on a transition with a thermally unpopulated lower level. It also was apparent that $CaF_2:U^{3+}$ strongly absorbed in the visible and would thus be spectrally matched to a high pressure xenon arc lamp. The other system, studied by the same people, was divalent samarium. Its spectrum looked very promising also, provided the crystal was cooled to cryogenic temperatures.

L&A: Did uranium require cooling?

Sorokin: Yes, to at least [liquid] nitrogen temperature to take full advantage of the four-level scheme. Samarium absolutely requires cooling to 20K. To try these systems, we had to find somebody to grow the crystals under reducing conditions, because both uranium and samarium normally have higher valences.

Isomet [then] in New Jersey said they would make a run and dope calcium fluoride with samarium. We asked Walter Hargreaves at Optovac in North Brookfield MA to grow some calcium fluoride boules doped with uranium. He probably thought we were crazy, but he said he would do it if that's what we wanted, but we had to provide the uranium. So we sent him some rods, and got a phone call, "How do I cut pieces small enough to put in crucibles?" I said, "Use your native Yankee ingenuity!" Later, when I asked him how he solved the problem, he said, "I used an ax."

When we received the first shipments, I was thrilled because the uranium doped crystals were beautiful ruby red, and Isomet's samarium crystals were beautiful dark green. You know you've got the right valence when the crystals are strongly colored. Then we sent the crystals to Karl Lambrecht in Chicago to polish into rectangular shapes. It was just at this point that I heard on the radio that somebody from Hughes had announced he had an optical maser.

L&A: Were you surprised that the ruby laser worked?

Sorokin: Yes. I don't think that anyone thought it would work that quickly. I believed Ted Maiman's results because he had done good work with microwave masers. But we were astounded at how he pumped it. He clearly must have had megawatts of light power, and we were thinking in terms of pump power of a few watts.

Stevenson called up Schawlow, whom he knew, and Schawlow told him about the photographer's flashlamp Maiman had used for an optical pump. Stevenson ordered one right away. We decided to forget about the polished square approach, because all this pump power was now suddenly available. Besides, we figured that our crystals would lase at a thousand times less pump power than ruby because of the four-level scheme. So we had Karl Lambrecht cut cylinders out of some of our other boules, and we had the ends silvered. We put the uranium crystal in a dewar equipped with an optical port, and set the flashlamp adjacent to this port. We figured that this somewhat inefficient optical coupling arrangement should still work, because the threshold for lasing of our crystal should be so much lower than that for ruby. The first time we tried the experiment, in November, 1960, it did work, and we saw the oscilloscope trace go off scale. Samarium worked for us the same way when we tried it a few weeks later, right at the beginning of 1961. These were the second and third lasers on record.

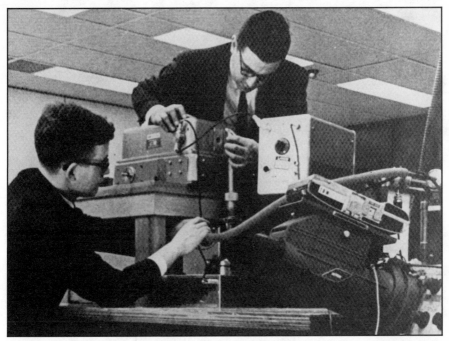

Sorokin and colleague Stevenson adjust the $CaF_2:U^{3+}$ laser at their IBM lab (circa 1960).

*"The first time
we tried
the experiment...
we saw the
oscilloscope go
off the scale."*

After succeeding at making uranium lase, we wrote an article for *Physical Review Letters*. Stevenson, being direct and aggressive, said, "We're not going to send it. We're going to drive down to Brookhaven and tell Sam Goudsmit [the editor] we want a decision before we leave." I said, "Mirek, you can't do that." "Nope, we're going to do that." So we got into a car and went to see Goudsmit. He was slightly confused about the differences between masers and lasers, and said he didn't want another "maser" paper.

L&A: That was how Maiman's ruby-laser paper was rejected.

Sorokin: We were aware of that. Mirek talked Goudsmit into accepting it, and as we were leaving, Goudsmit said, "Next time, tell your people from IBM not to come down here with machine guns."

L&A: How did you get involved with the dye laser?

Sorokin: Stevenson had been running a mutual fund on the side, and when top management told him he had to pick between IBM and the mutual fund, he left IBM. In 1964 I was working with John Lankard, whom I had hired as a technician in 1960. We started to try to develop a passive Q switch, which we thought we could do with an absorber that bleached at high light intensities. When I began looking for a substance with strong absorbance at the ruby wavelength, I quickly found that organic dyes had huge oscillator strengths and were highly absorbing in the visible. In particular, I found that phthalocyanines— complexes with metal ions at their centers—all absorbed near 694.3 nanometers. Chloro-aluminum phthalocyanine looked best, so we asked a colleague of ours, John Luzzi, to synthesize some. John was a very generous person; he didn't make us a gram, he made us a whole pound. That was to be important later, because it actually led to the discovery of the dye laser.

We placed these phthalocyanine solutions right in the cavity of our ruby laser. We fired the laser and, sure enough, instead of the usual train of spikes, out came a 20-nanosecond giant pulse. The thing worked so simply, so well, and it was so durable.

The spectral properties of the phthalocyanines were quite striking. They luminesced if you picked the right metal ion; thus, stimulated emission was one possibility. Another possibility was that one could get resonantly enhanced stimulated Raman emission, which actually wouldn't have been particularly noteworthy, but which seemed interesting to me at the time. Chloro-aluminum phthalocyanine was unusual in that it was quite soluble in ethyl alcohol, but its absorption peak in this solvent was shifted somewhat from 694.3 nm, so we had never tried this particular combination as a Q switch solution. I thought it might give the Raman effect, and at about 4:30 pm on Friday, Feb 4, 1966 we zapped it with our big Korad ruby laser and took a spectrum of the scattered light with our trusty old spectrograph. The plate had a black smudge, so we knew we had something, but I had to leave.

I thought about it over the weekend, and Monday I told Jack, "We have to take another shot, only let's align some mirrors with the cell." We did this, then fired the ruby laser. Jack came back from developing the plate with a big grin on his face. There was one place in the plate that the emulsion was actually burnt. There was so much light we knew we had something. It was not stimulated Raman emission because it was right at the peak of the dye fluorescence. It was laser action in the dye.

L&A: What did you do with the dye laser after that?

Sorokin: Initially we pumped the laser transversely. Bill Culver, who was then at IBM Federal Systems Division, suggested end pumping. It worked, and we got a beautiful beam. Then we decided to place an absorber in the cavity. It wasn't clear that we could make a Q switch, because there was basically no energy storage in the dye, but we wanted to see what would happen if we added a polymethine dye that absorbed at the wavelength of the phthalocyanine dye laser. Ernest Hammond, a summer faculty visitor in our lab, took the absorption spectrum of the polymethine dye we had selected. He told me that the dye also fluoresced very strongly. I asked if it absorbed at the ruby wavelength, and he said it did. So we said, "Let's try that. Maybe that works as a laser, too." We tried it and it worked, so we began to see that the effect was pretty general. We got all the dyes we could get from the chemical supply houses, even from the chemical stockroom here. We got our ruby going on the second harmonic, and just began pouring dyes into cells and pumping them to try to make them lase. Many of them did. I remember one afternoon we went down the aisle here at our lab asking our colleagues, "What color do you want?"

I really missed the boat completely on one of the most important things about dyes: Their ability to produce tunable, monochromatic light. The persons who discovered that, but never seem to get much credit for their efforts, were Bernard Soffer and B.B. McFarland at

Hughes. In a brilliant experiment, with a diffraction grating set up as one of the mirrors, they showed that, not only did the dye laser spectrum collapse to a narrow line, but also that you could continuously tune it by rotating the grating. We just hadn't thought of it.

The next year, 1967, we published a long paper on our cumulative dye laser research up to that time. It was published in the *IBM Journal of Research & Development*, which was where we published the first dye laser paper. I liked working with the editor, Hunt Gwynne, who was interested enough in our work to insist that it be written up clearly and carefully. He read our article and said, "How about a conclusion? Any long article likes a nice conclusion." I thought and replied, "What about the possibility that the dyes could be flashlamp-pumped?" That was the trigger for our next experiments.

We had heard that some people at Hughes had tried pumping dyes with big flashlamps and that it hadn't worked. D.L. Stockman at General Electric, who had proposed flashlamp pumping of dye lasers even before our ruby pumped results had been achieved, had calculated that it should be possible. He built a complicated, fairly fast flashlamp that would have worked on most dyes, but not for the one he picked, perylene. He published a paper on his negative results and just left it there. I had learned about triplet-triplet absorption while preparing the conclusion for our long dye laser paper, and figured that a flashlamp with something like a microsecond duration could excite most dyes and get them lasing before the triplet state filled and killed the laser again.

Jack Lanckard and I saw a picture in *Physics Today* of a coaxial, disk-like capacitor made by Tobe Deutschmann in Massachusetts, and thought that if we combined that with a coaxial lamp, we should be able to make a very fast discharge. We bought two capacitors, a small one we called Baby Bear, and another one we called Papa Bear. We assembled Baby Bear in the coaxial geometry and saw a discharge and some fluorescence, but no sign of a laser. We figured that might happen, so we decided to try Papa Bear. We were slowly charging the capacitor up to its rated voltage of 20 kilovolts. (The lamp was designed to self-flash when the applied voltage exceeded the breakdown voltage of the air inside the coaxial tube.) The laser was aimed at the wall or something, and at the last minute, as we were charging it up, I moved over to the other side to get a better view. At exactly the moment I moved, the lamp self-flashed. I turned around and

Lankard said, "It worked. I saw the beam on the back of your shirt."
The flashlamp-pumped dye laser indeed worked fine, and again we went
down the aisle and asked what colors did people want to see, and we
were busy demonstrating the new laser for everyone all that afternoon.

L&A: I gather you didn't have to worry about top management coming
around and saying, "You can't do this"?

Sorokin: That's right. In fact, after the discovery of the dye laser they
made me an IBM Fellow, which has really given me complete in-
dependence. I'm here in Jim Wynne's group, but I can do whatever
project I feel like doing. I'm lucky in that regard.

L&A: What are you doing now?

Sorokin: In the early 1970's, we did a lot of four-wave mixing in metal
vapors, both to generate tunable infrared and to generate tunable VUV.
We also took an idea of Bob Byer's and generated 16-μm radiation
in parahydrogen in a mixing scheme that's very close to what Los
Alamos finally adopted [for molecular laser isotope enrichment in
uranium], but then the isotope separation project went to Livermore
instead. What we've studied recently is most applicable to laser induced
chemistry. We take molecular spectra of reactions initiated by laser
pulses. Everyone knows about CARS, coherent anti-Stokes Raman
spectroscopy, as a way to take transient Raman spectra. What we've
developed here over the past four or five years is a technique called
TRISP, standing for Time-Resolved Infrared Spectral Photography.
This is a way of taking time-resolved infrared spectra with the same
time resolution you would get in CARS, using 10-nanosecond lasers
thus far.

We can apply it to the study of fast chemical reactions. We've done
work in initiating thermal explosions of methyl isocyanide with a
carbon-dioxide laser and watched the isomerization to methyl cyanide
develop in time. We studied laser-initiated gas-phase explosions of
hydrozoic acid, again initiated with a CO_2 laser. We've studied the
photochemistry of chlorine dioxide. TRISP is a real workable technique, and we used it very recently to study transient excited-state
absorption of a molecule (DABCO) we were considering as a two-
photon laser candidate. We're pretty proud of TRISP, but so far there
are no other takers, although Dave Moore at Los Alamos is apparently
looking into it. Maybe that's because it's a little complicated. It's
somewhat harder to set up than CARS, and you need a heat-pipe oven
for the metal vapor, but we're fairly comfortable with it.

We have plans to develop TRISP into a subpicosecond technique. It would be very interesting to look at a molecule hit by a UV pulse while it's still in the process of photodissociating. Chemists talk a lot about mechanisms, but it's mostly in the form of hypotheses, deduced by observing what products are finally formed. If you could capture the dissociation process spectrally—and I think infrared is probably the best spectral region for this—it would be very exciting, but you do need subpicosecond time resolution.

I'm always on the lookout for a two-photon laser, which would emit pairs of photons. We looked at one promising candidate (DABCO) recently. After we did our homework, we found that the cross-section is a factor of five too small, which is, however, getting much closer than we've been before.

L&A: Is that the closest anybody's come?

Sorokin: Probably. I also think there is a useful anti-Stokes UV picosecond laser there [in DABCO]. The gain should be high enough. The main problem is that you need a picosecond infrared pulse at 2.5 μm to get out the energy you store in the vapor with a 1 J KrF laser pulse. We may actually go back and pick up development of my old 2.5-μm uranium laser for that, because nothing else is at just the right wavelength. If you start with, say, 10 millijoules in a picosecond infrared pulse, you should get 10 times more energy out in the ultraviolet.

L&A: What special advantages would a two-photon laser have?

Sorokin: A two-photon laser is so hard to build that I don't think it would occur naturally, as for instance does the maser. Somebody out there who has a two-photon laser and wants to let the universe know could send its beam out into space. Millions of light years away, when someone detects it, the photons will always come in pairs, showing that this is unusual light. There are some research possibilities; I think people are interested in the coherence properties. But I cannot honestly imagine another particular advantage of the two-photon laser.

L&A: What do you think of dye-laser technology today?

Sorokin: I never expected it to be this good. Right at the beginning, its continuous tunability and monochromaticity—Soffer and McFarland's result—was a tremendous surprise; I think that affected me more than any subsequent dye laser development. Then there was [Ben] Snavely's cw laser, and all the work of [Charles] Shank and [Erich] Ippen in modelocking the dye laser. I'm in great awe of the recent technique invented by Dan Grischkowsky and his co-workers to compress pulses from commercially available dye lasers down to a few tens of femtoseconds in a totally passive way with the use of optical fibers. I understand how it works, but I never would have expected it. The flashlamp-pumped dye laser hasn't been that much improved; it's still kind of a wild beast. But the demonstrated usefulness of dye lasers in spectroscopic experiments is just astounding.

L&A: If you had it to do all over again, would you still investigate lasers?

Sorokin: Oh yes. The field has been very kind to me, IBM's been very good to me, and I certainly feel that I've been very lucky. The real advantage I had was being right there when the field was starting. My own abilities are such that I'm better at doing the first-order investigation than the subsequent careful follow-up. Coming into science in the late 1950s, I was very fortunate to come into a field about to explode, where I could make contributions, and I've been well-rewarded, too. So the answer is a resounding yes.

ALI JAVAN

The Helium-Neon Laser

Ali Javan, a native of Iran, came to the United States to attend Columbia University, where he received a PhD in 1954 for microwave spectroscopy work performed under Charles H. Townes. After four more years at Columbia as a research associate and instructor, he joined Bell Telephone Laboratories in 1958. At Bell Labs he continued theoretical and experimental studies on gas discharges and laser physics.

This research culminated in the December, 1960 demonstration of the first gas laser, a helium-neon type emitting at 1.15 micrometers, by Javan, William R. Bennett Jr., and Donald R. Herriott. Soon after, Javan joined the faculty of the Massachusetts Institute of Technology, where he now is Francis Wright Davis Professor of Physics. He founded and headed the MIT Optical and Infrared Laser Laboratory. He is also the founder and chairman of Laser Science Inc. in Cambridge, Mass.

Javan received the Franklin Institute's Stuart Ballantine Medal in 1962 and the Optical Society of America's Frederick Ives Medal in 1975; he is a fellow of the National Academy of Sciences, the American Academy of Arts and Sciences, the American Physical Society, and the OSA.

Jeff Hecht conducted this interview on February 28, 1985, in Javan's office at Laser Science Inc.

L&A: How did you first get involved with lasers?
Javan: Science has been in my blood since I was a little kid, and I was always fascinated by light and radiative processes, so I got into physics. I did my PhD thesis on microwave spectroscopy at Columbia University under Charlie Townes. I was not working on the microwave maser then,

but I could see the whole field evolving. Those were exciting days at Columbia, with Charlie, [I.I.] Rabi, and Willis Lamb discovering masers, the electron anomalous g-factor, and the Lamb shift.

After my thesis, I became more involved with masers when Charlie was on a year's sabbatical leave in Paris. My interest led to my discovery of the idea of a three level maser, in fact long before a group from Bell Labs published the first experimental work on the subject; actually, about a year before. I delayed publication because I became fascinated by a coherent, two-quantum effect, similar to a Raman process, which takes place in the three level maser. That, in turn, led me to the discovery that a Stokes-shifted Raman transition can produce amplification without requiring a population inversion. I published my three level maser theory in 1957 and the Raman work in a French journal in 1958.

Charlie Townes was encouraging me then to think about sub-millimeters and microwaves because we had resonators at these long wavelengths for feedback. Although I was aware of the possibility of achieving gain at short wavelengths, I did not know how to make a laser because I was not aware that a two-mirror Fabry-Perot optical resonator could be used to introduce the feedback.

About this time, Sid Millman arranged an interview for a job at Bell Labs and I talked with Art Schawlow there; he told me about the idea he and Charlie were exploring to make an optical resonator. I have always had a close relationship with Charlie, and I am really very grateful he did not talk to me about his ideas on lasers because that way our ideas did not get mixed. I was able to develop my own thinking.

After I learned that a Fabry-Perot might induce optical feedback, I rushed back to Columbia and started to calculate the same afternoon. I never really read the Townes-Schawlow paper [in *Physical Review*] because, by the time their paper came out, I was already deeply involved in my gas laser work. Even before I started at Bell Labs, I realized that, for a gas laser, you needed to find a pumping mechanism better than optical pumping. I should underline that I always looked to gaseous media. I am a gas-laser person; I don't do solids. I prefer the simple interactions of single atoms or single molecules with a radiation field. The Townes-Schawlow paper primarily related to cesium and some other species pumped by an incoherent light source. I started looking into other possibilities, and that led to my discovery of gas discharges as media where non-equilibrium conditions could lead to

Ali Javan

the presence of inverted populations.

I went to Bell Labs in 1958. There a dilemma evolved. RCA had previously inspected my notebooks on the three level maser and had established that my dates preceeded the Bell dates. They had paid me $1,000 for patent rights and were contesting Bell Labs' application. For the first six months or so working at Bell, I was dealing with RCA *and* Bell patent attorneys; it was very uncomfortable. Luckily, RCA did some marketing studies and concluded that the maser amplifier was not commercially viable. Thank God for that. We agreed to drop it and let Bell have it.

When I got to Bell, they had just dismantled their gas discharge department. My boss, Ted Geballe, told me to go buy a Varian magnet. I told him I didn't know what to do with it, but he told me to buy one anyway because sooner or later I would be needing it to study solids. So I ordered one and didn't even use it once. I published my initial proposal for producing population inversions in gas discharges in *Physics Review Letters* in 1959, and went after discharges on my own.

L&A: A lot of the other early lasers, both gas and solid-state, have just faded away, but helium-neon is still widely used. How did you manage to make such a successful choice so early?

Javan: It was no accident that the first gas laser is the largest-selling gas laser, with about 250,000 a year produced now, worth maybe $40 to $50 million a year just as components. I made a careful selection of a system with the promise of being the best medium for the first laser, and that best medium has stayed with us until now.

Helium-neon was one of the cleanest systems I could find. It also was a medium where I could show there was gain without first having to make the laser. Even after I had convinced myself that helium-neon was the best gas medium, there were a lot of non-believers telling me that gas discharges were too chaotic. They said there were a lot of uncertainties, and I had nothing I could control.

L&A: How did you go about proving them wrong and building the laser?

Javan: Nowadays if you have an idea of a gas system that might lase, you can align two mirrors two or three meters apart, put in the gas, and see what comes out. I couldn't do that because I didn't have a laser to align the two mirrors.

I decided to first establish that I had gain in the tube, then to try to align the mirrors by trial and error. That is how it worked. Without all my preliminary work to preset the HeNe discharge at a known gain, it would have been impossible to make it work. You could not have varied that many parameters and kept wiggling the mirrors without knowing that you have gain.

L&A: You started out working alone. When did others join you?

Javan: I always enjoyed working with others; otherwise this business would be too lonely. After I had done a lot of the early work, I persuaded Bill Bennett to come to Bell Labs for a year or two to join me. He had been a friend at Columbia, and was on the faculty at Yale University. He was very helpful, and I remember working late nights with him for over a year.

I also worked with Ed Ballik, who came to me as a technician, but was really a high-level person who made basic contributions. You don't hear his name much, but he was a very important part of the team. Later he got a degree at Oxford University, and now he's teaching in Canada.

The initial impetus to get Don Herriott came from Al Clogston and some other people at Bell, because he had been in optics before. It was essentially the four of us, with the support of some other people at Bell Labs.

L&A: How did you get from observing gain to having a laser?

Javan: I had the gain six or eight months before I had the laser working. I could see the gain, but I had non-believers. I had measured the excitation transfer from helium to neon, but people wouldn't believe that I had, in fact, measured the right cross-section. A year later, when I had the final measurement, that initial transfer cross-section turned out to be exactly right. But the non-believers said the gain I saw was really some nonlinearity in my detection system.

That really got me going. "By God," I said, "I'm going to make the damn thing work." I decided that the gain was going to be there. A lot more went into it, but the system finally worked on December 12, 1960; I remember it was 4:20 in the afternoon and it was snowing.

One interesting thing happened just after it worked. About six months earlier Ed Ballik had brought in a bottle of wine that was a hundred years old. We kept it to open after the laser worked. A few days later, I called the head of Bell Labs and invited him to come have this hundred-year-old wine. He said he would be very glad to come, then said, "Oh, oh, Ali. We have a problem!" This was two or three in the afternoon. He wouldn't tell me what the problem was, but said he would come at 5:30. Later that afternoon, a memo was circulated through the lab. It turned out that some months earlier they had disallowed liquor on the premises. The new memo stated no liquor was permitted unless it was over 100 years old. After that, he came.

After it worked, my friends in the Bell Labs administration finally told me that other people in the administration had called my efforts a wild-goose chase. In fact, they had been talking of cutting off my crazy idea in three or four months. That made me shiver.

L&A: I have heard of other people having such problems. Hughes' management had told Ted Maiman to stop working on the ruby laser, but he went ahead anyhow. Do you think you might have gotten into that situation?

Javan: I do not think so. Even if it hadn't worked when it did, I am sure I could have convinced people at Bell to keep it going. It would be too dramatic to say that I would have been cut off. My friends in the administration told me with pride that they defended my project. Bell labs had made a commitment; they told me the whole thing cost over $2 million. I am not so sure about the figure; maybe it included the salaries of all the top brass of Bell Labs. I was spending a lot of money, though, and if it would have gone another year and the damn thing wouldn't have worked, then there would have been real problems.

L&A: Your first helium-neon laser operated at 1.5 micrometers in the near-infrared. Why not the 632.8-nanometer red line?

Javan: I was aware of other transitions, but I picked the 1.15-μm line because of external restrictions. I could detect longer wavelengths with a photomultiplier, and the gain at the shorter wavelengths was smaller according to my calculations. My study showed that 1.15 μm was where you had the best chance to get the highest gain and the best experimental conditions to show the gain and optimize it. Also, the branching spectrum from the neon fine-structure was better resolved than on other transitions. So we could do the spectroscopy to show we could get the gain above the 7% or 8% needed for the laser with 99% reflective mirrors.

If we had set up the systems for the red line, the helium-neon laser would not have worked then. To get high gain on the red line, you have to have tubing with such a small bore that you cannot see through it to align the mirrors with a conventional auto-collimator. Remember,

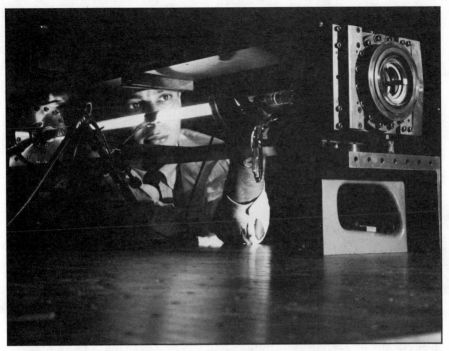

Javan demonstrates the first HeNe laser at Bell Labs.

"I am a gas laser person; I don't do solids. I prefer the simple interactions of single atoms or single molecules with a radiation field."

in those days we were working only with flat mirrors. Now we use curved mirrors, which are easier, and we have other lasers we can use to line up the mirrors.

L&A: Was it fairly straightforward to take the step from the infrared line to the red?

Javan: Everybody who made a contribution beyond the first set of laser transitions played an important role. Still, the red transition followed naturally and logically. The laser could be aligned already on the 1.15 line, and there was a lot known about the whole process. Later hundreds of transitions were discovered in HeNe lasers. One, at 3.39 μm, has such high gain it would even oscillate with the reflection from the back of your fingernail. All hell broke loose, but by then MIT had offered me a job. I really just passed through Bell Labs and dropped the helium-neon laser when I left. Bell Labs had done a very wise thing and encouraged other people to get into it within a month after my laser worked. The gas discharge department was re-established overnight.

L&A: The neon fluorescent tube was known long before anyone thought of the laser, yet it didn't seem to lead anywhere. Do you have any ideas why?

Javan: If you look back on the evolution of quantum mechanics and spectroscopy, you find that gas lasers, maybe a helium-neon laser, should have been discovered in the early 1930s. It shows that, even if the time is right, something may be totally missed. The German spectroscopists were studying gas discharge media and would plot the population of excited states versus discharge current. In the early 30's they studied neon extensively. A little extrapolation of their work would have given rise to an inverted population. By that time physicists knew the expression for negative dispersion, and there was a German paper with the subtitle "negative dispersion in neon." But at that time, when people experimentally saw a population with nonthermal distribution, they would immediately try to bring it back to thermal equilibrium. That meant no gain and no lasers.

Why was that? They were fascinated with thermal equilibrium because that led to Planck's discovery of black-body radiation law and the Einstein A-coefficient, the cornerstones of the discovery of quantum mechanics. This fascination with systems in thermal equilibrium inhibited workers from thinking you could have nonthermal distribution, which could lead to gain. In fact, more of our universe is in non-thermal states than is in thermal states. Now at that point, lasers would have been possible only in gases because solid lasers would have required developments in crystal growing that only came after World War II.

There is also the question of feedback, but if you look in old optics books, you will find Schroedinger's interferometer, an experiment from the early 1930s. He took a fluorescent crystal, polished its two ends, made them parallel and reflecting, and illuminated it with a flash of light. He was looking for dipolar and quadrupolar radiation inside the resonant cavity. Fabry-Perot cavities were known from the days of Mr. Fabry and Mr. Perot. So gas laser could have been discovered in the early 1930s, but by the mid-thirties people had presumably discovered everything they wanted to know about gaseous media and had moved to something else.

Suppose that gas lasers, including carbon-dioxide lasers, had been discovered before World War II. Then laser radar, not microwave radar, would have been the name of the game in World War II; microwaves would have come later. We would have now been working on microwave sources and radar. But because people missed the chance in the 1930s, we had to wait for Charlie and the maser, and then we were into lasers.

L&A: What took you from Bell Labs to MIT?

Javan: I always wanted an academic career. MIT made me an offer. Charlie Townes had just joined MIT, and I'm sure he had a lot to do with it. I set up my research lab, which I built up to a very large operation. For a year after I got to MIT, I had a joint appointment with Bell Labs and kept a lab there too. The distance from Murray Hill to Boston made me drop that very nice relationship later.

L&A: What sort of things did you do at your MIT lab?

Javan: I decided that I wanted to understand laser processes and do things with the laser. I have developed a whole lot of technology over the years. I did the original work in high-resolution laser spectroscopy. My original work on three level masers had turned out to be very basic in studying the interaction of an optical field with a pair of atomic resonances and branching to other resonances. The only new thing in optics is the presence of Doppler broadening in gases, which turns out to be quite a bit different than inhomogeneous broadening in solids. Willis Lamb's paper on the Lamb-dip was a very important contribution and revealed new features. In the 1960s, I introduced a number of ways to eliminate the Doppler broadening in atomic spectra and to obtain very high-Q resonances for very high-resolution spectroscopy. Later in the 1970s, I extended these to molecules in the IR. Now, with good dye lasers, a lot is being done in Doppler-free spectroscopy.

I take great pride in having originated the technology for extending microwave electronics into optics. This led to my high-speed diodes with a response time as short as 10^{-14} or 10^{-15} second. My work in this area led to the absolute measurement of light frequencies, in which the frequency of light is compared with a microwave clock. Now the National Bureau of Standards and others have followed my MIT work in their speed of light measurements. John Hall at NBS in Boulder has done a lot of beautiful work with considerable impact. My absolute wavelength measurement of Doppler-free lines in a CO_2 absorbing gas took me over eight years to do with my very talented students and colleagues.

I have unfinished business in the field I call "optical electronics," extending microwave electronics technology to the optical frequencies. I can envision oscillators and amplifiers at optical frequencies that would operate on the same principle as microwave oscillators. They would be classical devices in the same way that microwave or radio frequency oscillators are classical devices. These kind of elements wouldn't compete with lasers. They could serve as components for high-speed computer memories, in holographic imaging in real time, and in applications you couldn't even predict now. Such work requires access to special microelectronics which is now available. I have full intentions to follow this work at MIT.

L&A: What are you doing now at MIT?

Javan: I have never liked to work in research areas that are crowded with a lot of people doing a lot of good and bad work. So much of my work over the years has evolved into big fields. For this and other reasons, I am now in the process of changing the direction of my MIT research. In fact, I've given myself a little time to establish the next research phase.

My plans are to get into studies of the upper atmosphere with lasers. This is already a developing field, but I plan to do it in other ways—to extend quantum electronics into studies of upper-atmospheric phenomena, not just linear-type laser atmospheric remote sensing. I intend to take advantage of nonlinear interactions between laser radiation and atmospheric species in a variety of ways.

L&A: Is that what you're trying to do with Laser Science, the company you formed in Cambridge?

Javan: I did the planning for the company when I was on sabbatical leave a few years ago. My relation to LSI is, naturally, outside of my MIT activities and duties. The company is primarily an engineering operation based on the technology I have developed over the years. LSI is now introducing a whole new kind of laser technology that I know will have a lot of impact.

Most of the commercial gas lasers you can buy now were not just discovered, but were also perfected in the engineering sense, in the 1960s. A lot of things happened in the 1970s and 1980s, but where are they? What happened at the end of the 1960s is that the attention of the Department of Defense—which has been the prime force in pushing the engineering evolution of gas lasers—got shifted to high-energy lasers, mainly the CO_2 and the HF/DF lasers, which can be used to produce a lot of laser energy. You can be proud of the achievement in very high-energy gas lasers, but these very high energy lasers are not viable for commercial use. For that, you need lasers at medium or low energies.

Another reason that DoD shifted away from medium and low energy gas lasers in the 1970s was that neodymium-YAG was there. Now we find that YAG has great limitations for broad-based applications. Gas lasers can go deep into the ultraviolet and deep into the infrared; there are great varieties of them, and you can have precise control over the wavelength, even at a moderate pulse-energy. The medium and low energy range is where the action is, but nobody in the country has given it attention with respect to gas lasers. That's where Laser Science comes in—to develop medium and low energy gas laser technology for a variety of very large-scale DoD and commercial applications, where the market is now in the developmental stage.

When I mention low energy, I think of pulsed lasers producing 10 to 100 microjoules per pulse. If the pulses are in the nanosecond range, you can get 50 or 60 kilowatts peak power. You can do every nonlinear thing under the heavens with that much peak power. In particular, if you focus that energy to very small areas, you can get more than gigawatts power flux.

When I say medium energy, I mean a CO_2 laser, for instance, at energies of several joules and higher, with the possibility of obtaining very accurate and pure frequencies for coherent Doppler lidar and remote sensing. For that, the extraction of laser energy from an energetic medium at highly-controlled frequencies is the name of the game, and my company has the foremost expertise in this area and entirely new ways of doing it.

L&A: How do you feel about laser technology today? Has it brought any surprises?

Javan: Let's limit ourselves to gas lasers. Just a few months after the helium-neon laser, a doctor came to see me and started telling me about medical applications. That came as a kind of remote possibility and a surprise to me then. I just listened to him with great respect. The idea of focusing light and looking at photochemical reactions for medical use, and things like this recent work of cleaning arteries, they're just beautiful. If anybody had said he was working on a laser to make an impact on surgery, you would have said he was crazy. So in that sense there were plenty of surprises.

But then, surprise really is not the right word, because things happened gradually. I really had no idea of things like laser fusion or high-energy laser weapons in those early days; none of us did. But I was not really surprised because they evolved little by little, and I was with them as they evolved.

L&A: Where do you think gas lasers might go from here?

Javan: Before we think about new things, we have got to make the things we've already done in the laboratory really happen. To keep looking at the next thing and the next process is open-ended, though x-ray lasers and the like have to be done for the future. I think the technology of the 70s will happen and will have a large impact on industry and our very existence. I think the field is open for gas laser technology at medium and low powers. If you want to stick to a few milliwatts continuouswave at a fixed wavelength, it's forever helium-neon. But the new technology of making pulse intensities in the multi-kilowatt and higher ranges at nanosecond pulses available for use by the non-laser physicists, the biologists, and the chemists, and making them low-cost like a flashlamp, opens up large-scale commercial possibilities.

There are more of these things to come because the gas medium offers wavelength flexibility and the possibility of tuning. Laser radar has applications for peak power in the megawatt range extracted at a controlled frequency. The challenge is to extract that energy at a highly pure and stable frequency. You don't need a new laser—the

CO_2 laser is a beautiful one. The challenge before us is to make what we need happen.

In terms of the future and beyond, making an x-ray laser is fine, but we should also look for other ways of making coherent x-rays, other than a laser process which requires dumping a lot of energy into the medium. And I'm not talking about harmonic generation. There are other ways of extracting coherent continuouswave energy that don't involve inverted population and gain or a parametric effect. I have some ideas on that line. The future is open.

ROBERT N. HALL

The Semiconductor Laser

Robert N. Hall has spent his entire career in semiconductor research at the General Electric Research & Development Center in Schenectady, N.Y., which he joined after completing a doctorate in nuclear physics at the California Institute of Technology in 1948. In 1962, he conceived of a way to obtain laser emission from a semiconductor junction and assembled a team at GE which demonstrated the first gallium-arsenide injection lasers within a few months. Almost simultaneously a group at the IBM T.J. Watson Research Center in Yorktown Heights, N.Y., reported stimulated emission from GaAs devices that were similar, but lacked the Fabry-Perot cavity used for mode selection at GE. In a matter of weeks, a third independent group at MIT's Lincoln Laboratory also reported GaAs diode lasers. Those early devices required cryogenic cooling even for pulsed operation. Continuouswave, room-temperature operation of a diode laser was not demonstrated until 1970, long after Hall had turned his attention to other semiconductor research.

A Coolidge Fellow at GE, Hall is a fellow of the American Physical Society and the Institute of Electrical and Electronics Engineers; he received the David Sarnoff Award from IEEE in 1963. Jeff Hecht conducted this interview on October 11, 1984, in Hall's office at General Electric Research & Development Center, Schenectady, N.Y.

L&A: How did you first get interested in science?
Hall: When I was young, we lived in Puerto Rico. We were somewhat isolated, and I needed things to do. We had a set of the *Book of Knowledge* which had a section on "Things to make and things to

do." We did a lot of those things; I remember learning about making gunpowder and making a cannon and blowing various things up. The real start was when I came to live with my grandparents in New Haven for junior high school. My uncle, who was an inventor, took me to a science and technology fair in New Haven, and I saw a lot of things that just astonished me. I saw ball bearings bouncing off steel plates through hoops, and a tin-can motor that was just spinning all by itself, though there was a three-phase rotating field under the table. I *had* to learn how to do that. I got interested in motors, went to the library, and learned how to make a little motor with a commutator that actually worked. So my interest started with that science fair and maybe a natural curiosity. Certainly a big factor in getting me started was nudging from my uncle, who helped me to understand Ohm's law, and explained things, and showed me where books were in the library.

L&A: Where did you go to school?

Hall: I went to high school in Alameda, Calif.—then I went to Caltech. I graduated in 1942, and came to General Electric.

L&A: What were you originally doing here at GE?

Hall: I came here on a test program. I started in the research laboratory, and technical staff was very short—it was early in the war—so they asked me to stay on. It was a good place to work, so I stayed, working on continuouswave magnetrons for jammers and other microwave components. I also did a little work on special germanium diodes which Harper North was developing here. They were intended for microwave mixers, but he found that he could make very high voltage point-contact rectifiers.

L&A: That must have been very early in semiconductor research.

Hall: It was before anybody ever made a transistor. People were just playing with germanium, silicon, and other semiconductors, mostly for detectors and mixers for radios.

L&A: Then after the war you went back to graduate school?

Hall: Yes, I went back to Caltech and got a doctorate in nuclear physics. I came back to GE because I liked the environment and the people. Shortly afterwards, the transistor was announced by Bell Labs. I was talked into looking into it a little bit and found it very interesting. This was in the early days of semiconductors, and it was a very fruitful field.

L&A: How did that lead you to the semiconductor laser?

Hall: That is quite a long story. There had been some semiconductor work here, and it seemed to me that the principal problem was in

Robert N. Hall

getting sufficiently pure germanium. So I devoted a year or two to making very high purity crystals by fractional crystallization, which turned out to be very efficient. We got into pulling single crystals of germanium following the work at Bell Labs. We found we could make rectifiers by alloying and regrowth, and this gave us a way of making a *p-i-n* rectifier, which has become very well established as a high-power device. We worked on an alloying process for making transistors, then came a period when we were on tunnel diodes because Leo Esaki had announced this negative-resistance phenomenon from extremely heavily doped junctions. We explored tunnel diodes in germanium and silicon, and in III-V compounds. This fortunately gave us a good handle on making very heavily doped III-V compounds, gallium arsenide in particular. So we had in-house people—myself and fellows working with me—who knew how to build things from gallium arsenide with very heavily doped junctions. These are the same process steps you need to make a semiconductor laser. So we were sort of ready to invent the semiconductor laser once we thought about it.

L&A: At the time, weren't there a number of people talking about electroluminescence and III-V compounds?

Hall: Most of the III-Vs are direct-transition semiconductors, so radiative recombination is much more efficient than it is in germanium and silicon. The ruby laser and the gas laser had just been made, and lasers were an exciting field. People had written some articles speculating on ways to make semiconductor lasers.

L&A: Who were some of those people?

Hall: [Nikolai] Basov in Russia had done a lot of early theoretical papers, and Benjamin Lax at MIT. Maurice Bernard in France had done some very interesting work. It explained in terms that I could understand what was meant by a population inversion in a semiconductor. I was a stranger to lasers and first had to learn some of the principles. I remember that people had occasionally asked me whether a semiconductor laser was possible. I originally pooh-poohed the idea, but those questions made me look into it and learn what the requirements might be. I remember looking through some conference proceedings and thinking that it didn't seem those ideas would work.

L&A: What changed your mind?

Hall: The thing that really set it off was going to a device conference and hearing a paper by [R.J.] Keyes and [T.M.] Quist [of MIT Lincoln Laboratory]. I believe there was also a similar paper by Jack Pankove from RCA. They talked about very high intensity radiation from a GaAs *p-n* junction, which was really astonishing. At high current densities, they claimed that something on the order of a kilowatt per square centimeter of light was coming from those junctions.

L&A: Was that the first LED?

Hall: No, there were LEDs being made, but this was the first indication that they could be very efficient. Previously we thought maybe 0.01% efficiency might be reasonable, and here they are talking close to 100%. That really shook me up; that was the light bulb turning on. If things are really this efficient, what can you do with it? Now those thoughts about making semiconductor lasers began to make sense. With this evidence of very high efficiency and high power densities, I began putting numbers down. I remember on the train ride back from the meeting thinking about this and trying to write down numbers. It looked as though you could get a population inversion, and then the thing ought to work, if you could build the right kind of structure. So I began thinking how you might make a junction, include a cavity somehow, and get the Fabry-Perot geometry that would sort out the single mode you wanted and get coherent light coming out.

L&A: That is what you brought home?

Hall: Yes. I did some more analytical work, putting numbers together to see how they turned out. They looked very encouraging, so I thought about possible structures and worked out the idea of making a *p-n* junction, cutting it up, and polishing. I had been an amateur astronomer and back in high school had made my own telescope, so I knew about ways of handling optics and polishing so you could make little structures suitable for a laser. Nowadays you cleave lasers, but we did not know about that then.

L&A: How did the project get started?

Hall: We had a good deal of freedom to move from one project to another. I figured we would need half a dozen or so people and talked to a number of people who I thought might be able to contribute. They sounded interested, so I went to my boss, Roy Apker, and told him that I had some fellows rounded up and thought the idea might work, and asked how about having a go at it. He gave us the okay, so we went to work.

We had a very enthusiastic crew. Ted Soltys did the fabricating; he had been making GaAs tunnel diodes, and I suggested ways for him to make the junctions, so he got to work. Gunther Fenner was very good at electronics, so he began setting up equipment to pulse the diodes. We knew we had to get as high a current density as possible, and that to drive them that hard without burning them out, you had to give a quick pulse, then look for any light coming out.

We did not know what might come out, so we thought of various tests, and got an imaging tube to convert the infrared GaAs radiation into the visible so we could see the radiation pattern and watch for some kind of change when the emission went coherent. We thought we would see a change in the current-voltage characteristic, but that never materialized. Jack Kingsley had been working on lasers and was familiar with the use of spectrometers to look for line narrowing. Dick Carlson also helped us with some material preparation.

We expected that our first experiments probably wouldn't work. We tried what we thought might work, but meanwhile we thought we would have to do some more serious basic work, so Carlson was working on materials problems associated with preparing heavily doped junctions. We made some junctions, got them assembled, and Fenner had his apparatus going to test them. He came in weekends sometimes, and one Sunday afternoon he gave Roy Apker a frantic call saying there was something going on that he did not understand, but that something was happening! We all got excited, and on Monday we all got together and looked at it. We were never able to reproduce the original thing he saw, a strange bar of light across a Polaroid of the image tube. But by Monday he had some things that looked more sensible, and we could see clearly that there were some very interesting patterns. Above a certain threshold current, the spectrum changed very drastically and you could see intensity patterns and modes showing up.

We knew we had something going for us then. It was a big rush from that point on. Some diodes worked, but of course most didn't. Some did very strange things that we couldn't make much sense out of, but a few behaved in ways that we could understand and interpret as clear evidence for coherent light emission. Jack Kingsley got his spectrometer going and saw very clear spectral narrowing and multiple modes at the Fabry-Perot frequency spacing that we expected. We could see radiation patterns consistent with the number of lateral modes across the face of the laser, and taking into account the dimensions and wavelength of light, we could understand the far-field pattern.

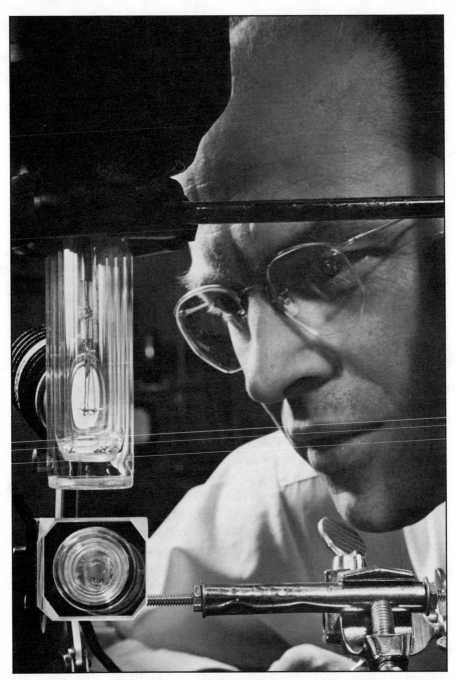

Hall inspects an early version of a GaAs diode laser mounted on a cryostat. The prism at the bottom of the photograph reveals a top view of the laser (circa 1963).

So we got enough to be convincing, and put it into a letter to *Physical Review Letters*. It came out on the same date the IBM group [M.W. Nathan, W.P. Dumke, C. Burns, F.H. Dill Jr., and G.J. Lasher] published their paper in *Applied Physics Letters*.

L&A: And just after that Lincoln Lab group [Quist, R.H. Rediker, and Keyes] had a paper in *Applied Physics Letters*. How long had you been working on semiconductor lasers?

Hall: It was a very short time. The device conference was in June, and the publication date was November 1. [The journal received the paper on September 24].

L&A: Had IBM and Lincoln Labs been working on the problem longer?

Hall: I believe so. They had been very interested in coherent emission and did some very worthwhile work on pointing out the importance of a direct-transition semiconductor, such as GaAs, for efficient conversion of electricity to light. I think they looked at the possibility of a laser but didn't describe how it could be done. I'm not sure if they were actually working toward a laser. [IBM did have a contract from the Army Signal Corps to work on a semiconductor laser—*Ed*] They were studying luminescence from GaAs diodes, pulsing them harder and harder. In a few cases they noticed spectral narrowing and knew they were getting stimulated emission, which is part-way toward coherent light from a laser. That was the result [the IBM group] published.

It wasn't the same structure we had; our laser had a Fabry-Perot cavity, which you could say was a proper structure aimed at making a semiconductor laser. They did not have anything that would give them mode selection, but they knew they had something very close to a laser. It wasn't long after that—I suspect after they saw our paper—that they recognized they needed a reflective cavity around it.

Another point is that most papers that had speculated about semiconductor lasers had indicated that the mirrors would be parallel to the junction plane, so the radiation would go perpendicular to the junction. The thing that made ours work, and the basis for all subsequent semiconductor lasers, was arranging the Fabry-Perot mirrors so the radiation would bounce back and forth in the junction plane. This gave a relatively long path for amplification. Also, right from the start I recognized that if the radiation was going perpendicular to the junction you would have an awful lot of losses from absorption by the heavily doped material on both sides of the junction.

L&A: Had you been aware of others working on semiconductor lasers?

Hall: We had to guess. We knew that Keyes, Quist, and Rediker were because they showed their work on high-efficiency luminescence. And we knew that Pankove probably was because he had given a similar paper. We knew that IBM had thought about the possibility but had seen nothing to indicate they were actively pursuing it. Their letter took us totally by surprise.

L&A: Did the others know about you?

Hall: No, I think we caught them totally by surprise. We had done tunnel-diode work, which in a way is a close step because it involves degenerately doped semiconductors and gallium arsenide. So we had all the ingredients but had never done anything on optical emission that I can recall.

L&A: Didn't you have some interaction with Bernard?

Hall: He came to our laboratory several times, discussing the possibility of semiconductor lasers. He appeared when we had one going but before we had submitted our paper. I felt a little bit awkward trying to discuss the problems of how you might go about it and what the problems were, and not being able to tell him that we had one in the next room.

L&A: Why do you think you were first?

Hall: The device conference provided an immediate stimulus to all of us. I guess we all had an equal chance then, and MIT and RCA probably had a head start. But we happened to have people with the right talents who could be put together and who weren't tied to other projects, and a manager who said go to it. We had a loose and informal organization, with people who knew how to work together and could do it very efficiently and quickly.

L&A: Then the environment at GE helped make it possible?

Hall: Yes. Also, we happened to recognize the necessary ingredients; some thoughtful design went into it to begin with.

L&A: Were you the manager or leader of this group?

Hall: I have never been a manager, but I guess I could be called the leader. It was mainly my idea, and the other fellows all pitched in and worked hard.

L&A: What were your hardest technical problems?

Hall: First, we had to get some good gallium arsenide, which was hard to find then, though we had an assortment of material on hand from the tunnel diodes. We had to cook up ways of diffusing junctions.

A lot of the GaAs was just dead, but in some cases we found junctions that would light up pretty well.

Another problem was making the structures themselves. I learned how to cut, lap, and polish these things, and to check the parallelism of the two sides, which was tricky to achieve. Ted Soltys mounted them on headers with good husky leads so they could handle high currents. Gunther Fenner set up the electronics to get very high current pulses and set up the liquid air flasks so you could look at the light coming out. It wasn't anything you would call a scientific challenge, but it was an engineering and technical challenge to assemble this equipment in a short time and get it running.

L&A: What more did you do with semiconductor lasers?

Hall: We followed the field for a while, trying to understand the mode patterns and trying to get more efficient lasers. Then we began to think about possible applications, which you could see once you learned something about their properties. They seemed to be mainly in communications, where GE had relatively little interest, so we didn't see major applications for semiconductor lasers within the company. Lots of other things did need attention, so after about a year or so most of us trickled off, and the semiconductor laser work just stopped.

L&A: Was that a management decision?

Hall: I wasn't under pressure to drop it. If there was a clear need in the company, operating components would be willing to pick it up, but we didn't have that kind of encouragement.

L&A: The need for cryogenic cooling obviously was a major drawback of early semiconductor lasers. Why did it take so long to get away from that?

Hall: You had to make very efficient heat sinks and very small structures so the heat could spread out laterally. In our early lasers, the junction was across the entire plane of the device. To go to higher temperatures, you need stripe lasers, so the current density will be very high just in a very narrow region where it is needed, and the dissipated power is as little as possible so the heat could spread out laterally and into the heat sink. They had to develop copper and diamond heat sinks and just engineer the dickens out of this problem of getting the heat out. It took quite a few years, with a lot of tough engineering steps.

L&A: What are you doing now?

Hall: I have been looking at problems relating to very large scale integrated circuits in silicon (VLSI), which GE is putting a big emphasis on. One of my long-term interests has been in semiconductor defects and possible impurity contamination, so I have been looking into some of our processing steps to see to what extent impurities are problems.

L&A: Are you surprised at how far semiconductor laser technology has come?

Hall: Yes, I am very much surprised, or maybe not so much surprised as impressed. Although I'm not actively involved with semiconductor lasers, I am familiar enough with the field to recognize that an awful lot of tough engineering problems have had to be worked out. It has been very interesting watching this develop through all the painful steps to the point where they have very efficient singlemode lasers working at room temperature, tailored to the right wavelength. It is just a marvel the way it has grown, and the way the various formidable problems have been chopped off, one at a time.

L&A: Where do you think the technology might go?

Hall: I suppose there will be lots of little gadgets, in cameras or home recorders or communications. I expect fiberoptic communications will crop up all over. I suspect that the semiconductor laser is so small it will crop up in applications you really can't predict. But I guess I have let that world go by. I do not so much follow the instrumentation end of things as I do try to understand the basic physics of semiconductors. There are plenty of mysteries down at the microscopic end of things to keep me occupied.

L&A: If you had it all to do over again, would you still work on the semiconductor laser?

Hall: I don't see any other way of answering that than yes. That was an experience I really treasure, a classical model for inventing something. You get some ideas, you put them together, the light bulb turns on and suddenly you see a way of doing it. You go at it as hard as you can, you get some guys working with you that you like to work with. The unexpected thing of this early success made it all the better. There are a few such things that happen in a career, but not many.

L&A: It is hard to find encores for that kind of thing. Do you regret having left semiconductor lasers?

Hall: No, in fact I feel a little bit, not quite ashamed of myself, but as though I had abandoned the field, and had all the fun of discovery without sweating out the hard work of, you might say, rearing the baby.

C. KUMAR N. PATEL

The Carbon-Dioxide Laser

A native of India, C. Kumar N. Patel came to the United States for graduate study in electrical engineering at Stanford University. After receiving his doctorate in 1961, he joined Bell Laboratories. He has become increasingly involved in management, and since 1981 has been executive director for research of the physics division at AT&T Bell Labs, Murray Hill, N.J.

Patel's pioneering work on molecular gas lasers included discovery of the 10-micrometer carbon-dioxide laser and the 5-μm carbon-monoxide laser, and demonstration of high-power output from CO_2. He also developed the spin-flip Raman laser, which he used to measure nitric oxide concentration in the stratosphere in pioneering experiments that played an important role in the debate over the fate of the super-sonic transport. Other research interests have included absorption measurements in highly transparent materials, spectroscopy of solid hydrogen, and medical uses of CO_2 lasers.

A member of the National Academy of Sciences and the National Academy of Engineering, Patel is also a foreign fellow of the Indian National Science Academy. He has received numerous awards, including the 1966 Adolph Lamb Medal from the Optical Society of America, the 1976 Zworkin Award from the National Academy of Engineering, the Institute of Electrical and Electronics Engineers' Lamme Medal in 1976, Texas Instrumentation Foundation's Founders' Prize in 1978, and OSA's Charles Hard Townes Medal in 1982.

Jeff Hecht conducted this interview on November 6, 1984, in Murray Hill, N.J.

L&A: How did you get involved with physics and lasers?

Patel: I did not start out to be a physicist. I had planned to go into the Indian Foreign Service after I got my bachelor's degree in engineering. However, you had to be 22 to take the entrance examination, and I was three years too young when I graduated. I wanted to do something useful during those three years, so I decided to get a PhD. After the first year at Stanford, I realized I was having so much fun doing research in physics that I wasn't going to go back and take the foreign service exam.

L&A: What was your doctoral research?

Patel: I studied ferrimagnetic resonance in yttrium iron garnet and worked on a narrowband microwave filter under Dean Watkins, who soon afterwards left Stanford to devote full time to the Watkins-Johnson Co. I finished up in early 1961 and came to Bell Labs. My supervisor asked what I wanted to do, the standard question you ask a new PhD at Bell Labs. I had had enough of microwaves, so I said I wanted to do high-resolution spectroscopy with lasers. I didn't know much about lasers or spectroscopy, but it sounded interesting. This fellow was quite savvy, and he suggested I start with lasers and see where that would go.

L&A: How aware of lasers were you then?

Patel: I had seen newspaper accounts of Ted Maiman's ruby laser, and read Ali Javan's paper on the helium-neon laser in *Physical Review Letters*. Arthur Schawlow had had an article in *Scientific American*. But that was about all. One virtue of this institution is that it lets people get into fields where they have no prior knowledge. All they require is enough basic understanding of science.

L&A: Then they basically turned you loose?

Patel: That's right. My supervisor pointed me to people like Ali Javan and Bill Bennett, to bounce ideas off. The large collection of people and accomplishments is one strength of this place. But another is that it's so big that it's impossible for a newcomer to talk to too many experts, because there are so many of them, and you don't know who they are.

The basic naivete of an individual who comes into a new field is a tremendous advantage in getting things done, because he does not know what some old fogeys have decided is impossible. I know how I decide now what is not possible today. The trouble is that two years from now I just remember the decision, not the arguments I used

C. Kumar N. Patel

to arrive at that decision. Things will have changed in two years, but I will remember my conclusion, not the caveats that went with it.

L&A: Did you have any particular ideas in laser spectroscopy?

Patel: I really couldn't have done any spectroscopy because there were no tunable lasers. There were just a few lasers then, mainly ruby and helium-neon, so the first thing to do was to try to make new types. We were being taught that helium-neon was an outstanding system because helium transferred its energy to neon through selective excitation and a population inversion. I wondered if you could get away from those two-gas systems. If you could get the inversion by an incoherent excitation mechanism such as a discharge, without needing an intermediate gas to transfer the excitation, then it seemed there would be an enormous number of possible laser systems.

The major breakthrough came within six months after I got here. I showed that you did not need helium in a helium-neon laser. It did not work quite as well as the helium-neon system, but it proved you did not have to have that kind of energy transfer. Of course, a few years later somebody showed that you could put almost anything in a discharge tube and get a laser if you hit it hard enough. That was disconcerting at the time.

L&A: Ron Waynant once called that the "telephone pole" theory of lasing, the idea that if you hit a telephone pole hard enough it would lase.

Patel: Schawlow and one of his students demonstrated that dramatically in the mid-sixties when they fired a ruby laser into a bowl of Jell-O. There was a bit of cheating in that it was doped with Rhodamine 6G, not the kind of Jell-O you would want to eat, but it was close enough. However, before we could get to that, we had to learn how to walk.

I had started looking for a tunable laser for spectroscopy. The next best thing was to have laser lines close enough to each other, so you could do point-by-point spectroscopy. I thought a large number of lines would let us do that.

I did have another approach to spectroscopy. One of the earliest high-resolution laser spectroscopy experiments was one I did using a xenon laser. I shifted the 2.02-micrometer wavelength by applying a magnetic field to the discharge, and was able to map out the Doppler profile of xenon in another tube. It wasn't tunable by a lot, but it was tunable enough to use. Before other tunable sources came along, some of my colleagues used magnetically tunable lasers for limited applications in spectroscopy.

This was about 1963, and I was getting much less excited about just looking for another transition. Gas lasers typically produced milliwatts of power. Someone here had built a 15-meter long helium-neon laser, which put out about 150 milliwatts, but that size was too much. A lot of people were saying that gas lasers were good as little laser lightbulbs for drawing straight lines, but that if you wanted high power you needed either ruby, neodymium-glass, or neodymium-YAG.

I changed my tack and asked if there might be something basically wrong in looking at rare gases. Of course, at room temperature only rare gases are atomic gases; everything else is molecular. The problem with atomic systems was that the first excited state is far above the ground state, so you have big losses.

I started looking at the possibility of using molecular systems. Carbon-dioxide was the first system I looked at seriously, and it had lots of advantages. You have to get away from electronic states, because they give you the same problem as in atomic gases—the levels are too far above the ground state. Diatomic molecules such as carbon monoxide have just a single ladder of vibrational states, and you can show from simple quantum mechanics that the vibrational level lifetime gets shorter for higher vibrational states. That isn't what you want in a laser candidate, so you have to go to at least a three-atom molecule. There you could have different vibrational ladders, and if you are lucky you can get a system where one vibrational mode has a longer lifetime than a lower-energy one.

Having done those calculations, it was quite clear that CO_2 could work. It did the first time we tried. You knew where to look for the radiation, near 10 micrometers, and by golly, it worked marvelously well. We got tens of milliwatts on the first shot.

L&A: Were you surprised that the CO_2 laser worked so readily?

Patel: Yes. I did not think I knew enough about CO_2 to have predicted it right the first time. Also, we were aware of Harry Boots' experiments in England, where most of the emission from molecular gases in a discharge was on atomic lines, so it wasn't entirely clear that you could keep the CO_2 molecules intact.

L&A: How big was your device?

Patel: A meter and a half long. This work made us realize that you did not need a complete inversion between two vibrational levels to

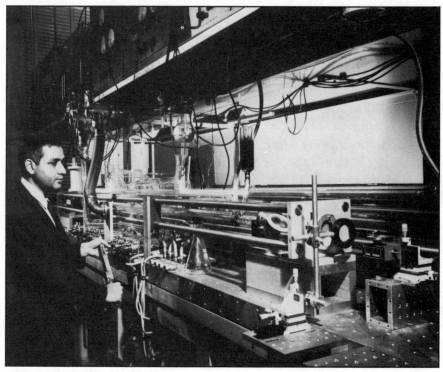

Patel adjusts one of his first laser accomplishments.

*"I try to stay away
from where most of
the people are,
because I think
the fun is most likely
to be where nobody
else has tried
anything."*

get lasing, because each vibrational level has a whole set of rotational states. Even when you have fewer molecules in the upper vibrational state than in the lower one, you still can have inversion between two rotational states.

The moment you realize that, everything becomes very clear. Then you say, why did I throw away the diatomic molecules? We went back and tried CO, and that worked too. I wouldn't have tried it first, because it didn't fit my preconceived notions of needing total inversion, but of course total inversion is not necessary.

Then I remembered about energy transfer in helium-neon, and wondered if there could be something like that in a molecular system. That led to the nitrogen carbon-dioxide laser system. Molecular nitrogen is very strange; if you throw energy into it, the energy just stays there, with a lifetime of seconds. Nitrogen's first excited vibrational energy level is the same as that of the CO_2 upper state. You put the two things together and the next day you get 10 watts from the same tube that gave you 10 mW before. I think that was the highest continuouswave power that anybody had had. Within six months we had destroyed all the earlier myths about where you could get large amounts of laser power.

By about mid-1964 it became clear that de-excitation wasn't really complete in carbon dioxide. Then the game was to find something else to add to the system to drop that lower level of CO_2. We tried water vapor, an idea that came from work on rocket engines. We also realized helium was good, because it has enormous heat conductivity at pressures of 10 or 20 torr. It turned out that the best mixture is carbon dioxide, nitrogen, and helium. One can add a few other things here or there, depending on whether or not you seal the tube, but beyond that I don't think there have been major advances in the gas mixture. By mid-1965 I had a 200-watt continuouswave CO_2 laser, which was more than enough power for anything you wanted to do in the laboratory.

L&A: Why did you move on from CO_2?

Patel: Gas laser work wasn't providing the same kind of excitement. I had had the field to myself until we published the paper on the 200-W laser, but then it began getting crowded. I try to stay away from where most of the people are, because I think the fun is most likely where nobody else has tried anything. Nonlinear optics was just becoming fashionable. Joe Giordmaine and R.C. Miller had just operated the first parametric oscillator. That gave me the germ of an idea for a tunable laser source using a carbon-dioxide laser. All we needed was nonlinear optics farther in the infrared than anybody had done before.

The whole question of nonlinear optics for the infrared was open. The first material I tried was tellurium, which turns out to have the largest known nonlinear coefficient in a phase matchable material. We stuck the sample in, and it generated the second harmonic of CO_2 exactly the way we predicted it would. The problem of making a parametric oscillator in tellurium turned out to be a lot more difficult than I had thought. Multiphoton excitation of electron-hole pairs in the semiconductor generates enough added absorption to overcome the gain.

About this time, 1966, Dick Slusher came to Bell Labs. He wanted to use the carbon-dioxide laser to study Raman scattering in semiconductors. His supervisor, Peter Wolff, had done a theoretical study that showed that Raman scattering from electrons in a semiconductor placed in a magnetic field could give you light output tunable in frequency. We got spontaneous light scattering the first time we tried it, using indium antimonide. The spontaneous Raman scattering was a very broad line, incoherent and very weak, coming out in all directions.

The money was in making a Raman laser, but the problem turned out to be very hard. Slusher got interested in other things, but I kept working on it. By this time Earl Shaw had joined my effort and eventually the spin-flip Raman laser lased in 1969. There had been Raman lasers before, but this was the first tunable Raman laser at any frequency.

L&A: What made it so hard?

Patel: We knew very little from theory or experiments about the nature of the Raman scattering process. We had fallen into a frame of mind that said we didn't have enough Raman scattering, and Raman scattering is proportional to the number of electrons, so let's increase the number of electrons. Things got worse, not better. It took me a year and a half to realize that if you have too many electrons, both the upper and lower levels of the Raman transition will get filled, and once they're both completely filled, you can't get stimulated Raman scattering. When we recognized that, the spin-flip Raman laser worked within two weeks.

L&A: So at last you had a tunable laser?

Patel: Yes, I had reached a point where spectroscopy was possible.

For a number of reasons, the indium-antimonide spin-flip Raman laser hasn't turned out to be practical, but we used it for five or six years as a workhorse for a variety of things. Soon after my work, Lincoln Laboratory pumped indium antimonide with a carbon-monoxide laser, which enhanced the cross-section because the pump frequency was close to the 5.3-μm bandgap. They also showed that the spin-flip Raman laser could emit continuouswave; our initial work was pulsed. I too jumped in and showed that going to high electron concentration InSb samples actually made the spin-flip Raman laser threshold go down because the spontaneous Raman scattering linewidth gets smaller. Also now you could use simple electromagnets or even permanent magnets for tuning the spin-flip Raman laser rather than the superconducting magnets used earlier.

We used spin-flip Raman laser spectroscopy for pollution detection, to see very small concentrations of gases. Lloyd Kreuger and I did the earliest tunable laser optoacoustic spectroscopy experiments. Optoacoustic spectroscopy was very useful because instead of trying to measure transmission through a sample, you try to measure the energy left behind. We showed you could measure absorption co-efficients as small as 10^{-10} inverse centimeters in a sample length of 10 centimeters.

I spent two or three years looking at a variety of pollutants including nitric oxide. Nitric oxide was rather interesting because there was tremendous sensitivity about what was coming out of automobile tailpipes, and nitric oxide was one of the constituents which was rather nasty. We showed that one measurement apparatus could measure concentrations from something around one part per billion to one part per thousand, implying a tremendous dynamic range.

The real excitement came when I decided that the right place for this kind of measurement was not on the ground. People were worrying about ozone depletion in the stratosphere, primarily from supersonic transports. Johnson at Berkeley had a model which said that if there were 400 SST flights every day, ozone concentration would drop significantly in two years, and this would increase the ultraviolet coming down. But nobody had measured the concentration of the nitric oxide that was supposed to catalyze the reaction.

We wanted to measure nitric oxide concentration at 28 kilometers. You can't take samples up there and bring them down, because by the time they get down, it will all be nitrogen dioxide. You want to put an apparatus up there before sunrise, and let it go through the whole day so you have an idea of the variation through a whole diurnal cycle. We and Sandia Laboratories built an apparatus and lifted it up with a balloon to 28 km. We found that the model had to be corrected by a small amount, and Johnson's calculations were basically verified by actual measurements. However, I am not sure if the nitric oxide problem in the end was part of the justification for scrapping the US SST program.

Having gotten to this stage, I had fulfilled in some sense what I wanted to do when I came to Bell Labs — high-resolution spectroscopy with lasers. It took me 14 years, but in the meantime lots of other things fell out of it. One of them was studies of hydrogen cyanide production from the platinum catalysts used to clean up automobile exhaust. Some of our work led the Environmental Protection Agency to rewrite catalyst specifications to mandate that the HCN coming from the tailpipe could not exceed some acceptable level.

L&A: Have you looked at other materials?

Patel: I have been looking at very transparent solids, mostly in the visible and the near infrared. We find there is no such thing as a totally transparent material; things absorb somewhere, and these small absorptions turn out to be important. We've been looking at the absorption coefficient of water, which people have done for at least a hundred years, and there is roughly a factor of 20 variation in reported data.

L&A: That large a variation for pure water?

Patel: As pure as can be. Whenever I find something like this, I consider the field ripe for doing something worthwhile. When you look at the problem carefully, you realize that the variations arise because the absorption coefficient is small, and you have to use very long paths if you use conventional measurement techniques. The optoacoustic technique lets us avoid seeing window absorption, and make measurements with absolute accuracy $\pm 10\%$. We also found the absorption minimum is at about 4700 Angstroms, not at 4900 or 5000 Angstroms where most people used to think it was. That is important for satellite-to-submarine laser communications.

L&A: Spectroscopy also is used in more research-oriented fields. Have you been working in those areas as well?

Patel: That's what I'm doing today, looking at solid molecular hydrogen. We don't think there is any immediate practical impact, at least today, but there is some really interesting physics. Most solids are atomic, but hydrogen is molecular solid. The hydrogen molecule sitting in a solid lattice still behaves as if it were a free hydrogen molecule. Because hydrogen atoms are so light, their vibrations cover much of the distance between two molecules, and the tremendous amount of overlap of wave functions between the molecules brings into focus lots of really exciting physics.

L&A: Aren't you also heavily involved in management?

Patel: Right now I don't get paid for doing research, I get paid for management. The research I do is what I call "for the soul." My boss never asks me how many papers I publish or what research I am doing. My getting into management is very similar to my having fallen into physics. I generally enjoy it and get a kick out of it. I don't think I would enjoy it if it were strictly what one might call managing people, but in the research area it is technical management, management of ideas.

L&A: How do you see your role?

Patel: In predicting the next breakthrough and putting our resources there. There is much more technical involvement than in managing other kinds of enterprises. I do very little of what is traditionally called management. I do no financial management — I have a specialist responsible for that. I do no personnel management; the specialist does that. My task is to make sure that the money that is entrusted to me by AT&T will create the needed breakthroughs, which means I have to find out where I am going to put my money.

L&A: The investment in the sort of laser research you did is now paying off in technology. Where do you think the CO_2 laser is going?

Patel: I see CO_2 lasers becoming an integral part of what I call a flexible manufacturing system. I have seen only one commercial CO_2 laser system which is what I call flexible, not a fixed laser waiting for parts to come by, but a thing mounted on a robot arm that can walk around the target, completely under software control. I expect to see that type of activity go up tremendously, as we begin to realize that we have to do something significantly better than the Japanese to stay competitive.

CO_2 lasers are also growing in surgery. One stumbling block has been the lack of a reasonable singlemode fiber that can carry 50 or 100 watts of power. Even without that, medical uses are clearly expanding. If somebody made an equivalent of silica fiber for 10 micrometers, it would take off.

L&A: What about laser technology in general?

Patel: There are many uses in what I call information technology. In communications, the laser is simply a light bulb, and the question is how fast you can turn it off and on. Beyond that, there is optical logic, and switching in a totally optical environment. Some people say optical switching is here and now, and there are devices that are called optical transistors, but I don't think we are there yet, no matter what one calls it.

L&A: What do you mean by that?

Patel: A transistor is a device in which a coherent or incoherent source of signal controls an incoherent source of power — what I call low-quality power. A switch that switches laser light with other laser light is not really a transistor, because the coherence and color of the light are more important features of what happens. That's not to take away from the work on bistable optical gates, but the point is that for a practical computer you immediately run into problems with power consumption. I would like to see some switching technology that lets you go optical to optical, so you only have to convert signals to electronic when a customer wants to put it on his terminal or pick up his phone. It has a long way to go, more than five years, but I think it will happen. An optical computer is even further out because we are up against materials problems that have no simple solutions.

L&A: If you had it to do over again, would you still work with lasers?

Patel: If I were starting today and forced to be a physicist, probably I would. But if I had a choice I probably would not do physics. There are fields such as biophysics and what people call "artificial intelligence" which are where physics was 20, 30, or 50 years ago. The fundamental understanding in those fields is very limited, so the possibility of making a significant advance for a unit of work is significantly higher than in lasers or physics.

L&A: Do you have any regrets about working with lasers?

Patel: Absolutely not. I may be stealing somebody else's words, but it has been a wonderful ride.

WILLIAM BRIDGES

The Ion Laser

William B. Bridges was born in Inglewood, Calif. in 1934 and received B.S., M.S., and Ph.D. degrees in electrical engineering from the University of California, Berkeley. He joined the technical staff at the Malibu Research Laboratories of the Hughes Aircraft Company in 1961.

After his first work at Hughes on microwave tube research, he was caught up in the excitement of laser technology. Bridges discovered the argon ion laser in 1964, after two years of research on helium-neon and xenon lasers. For the next several years, most of his efforts centered on developing practical ion lasers for airborne applications. Throughout the early 1970s, Bridges' efforts concentrated on various aspects of carbon-dioxide lasers and their suitability for military systems applications. Since 1977, he has been professor of electrical engineering and applied physics at the California Institute of Technology, Pasadena, Calif.

In 1983 he was appointed Carl F. Braun professor of engineering. Bridges is a member of the National Academy of Sciences and the National Academy of Engineering and is a Fellow of the Institute of Electrical and Electronics Engineers and the Optical Society of America.

Richard Cunningham conducted this interview on November 9, 1984 at Caltech.

L&A: How did you get involved in science in the first place, dating back to high school days or earlier?
Bridges: It seems like I have always been involved in science, although none of my family was involved. My mother worked for the local high school as a payroll clerk, and my dad was an auto mechanic.

I suppose the only influence in that regard was my grandfather, who gave me the full run of his workshop, and a great uncle who did the same. I was always interested in building things.

Actually, as you know, I am an electrical engineer. I really got involved in radio in the grammar school era by building crystal sets. By the time I was a freshman in high school I was an active radio amateur. I built my own transmitter and antennas and all the equipment needed. Radio electronics was a common route of entry for people in my era, when "electronics" connoted something you built rather than something you bought from Japan.

L&A: What were you doing professionally immediately prior to the ion laser work at Hughes?

Bridges: I had stayed on at the University of California, Berkeley to do my graduate research in microwave electronics. My thesis work was on noise in electronbeams of the kind used in traveling-wave tubes. I joined Hughes Research Laboratories in December, 1960, although I think I only worked two weeks and then took a leave to go back and finish my thesis. I returned to Hughes full-time in June, 1961, and joined the Electron Dynamics Department. My section was involved in the research and development of microwave tubes, particularly in low-noise microwave and millimeterwave tubes. This was a little more than a year after Ted Maiman demonstrated the first laser, so lasers were a hot item at Hughes.

My group was not involved in lasers specifically, at least not initially. But by about mid-1962 we had closed out an old low-noise tube contract. After seven years we couldn't think of any more to do, so we quit. Something you might not do these days is turn down money.

My supervisor then, Don Forster, was looking for some way to get into the laser excitement. In the early part of 1962 we thought of extending our microwave tube activity into very high-speed or broadband photodetectors. We started off by looking to see whether we could use use traveling-wave structures as photodetectors. This really followed some work that had been pioneered at Stanford by Tony Siegman and Burt McMurtry, his graduate student.

We actually built a couple of these things, and I worked cooperatively with Vik Evtuhov and Jim Neeland with their ruby laser setup. We used these high-speed detectors to puzzle out some of the funny mode properties of ruby lasers. Things were pretty crude in those days, and we weren't sure that ruby lasers were behaving as simple resonator

William Bridges

theory would predict. It turns out that it's okay and theory wins. You just need a better quality of ruby to get the expected mode properties. But my first work with lasers was really in conjunction with the microwave photodetector end of things.

Earlier that year, the helium-neon gas laser was first announced at Bell Laboratories, with the infrared line at 1.15 micrometers. At that point another group at Hughes started trying to build some helium-neon lasers following the Bell Labs design. We ordered one of their lasers for our phototube work. Well, the other group had some troubles, and one thing led to another. I got involved with this group, trying to get gas lasers from them.

About that time, the Bell people announced their red helium-neon laser in June, 1962. Mal Currie, who was then our associate lab director, said, "That's fantastic. Let's get into that." Overnight, I found myself in the gas laser business. The microwave phototubes were forgotten, and we were off and running with helium-neon lasers.

L&A: What influenced you to start working on the noble gas ion area?

Bridges: From mid-1962 we worked on helium-neon lasers. Remember, Hughes was not really a commercial manufacturer in those days. We were building tubes for our own internal use and were trying to find appropriate military applications. We got into some of the other noble gas neutral lasers, particularly xenon. In fact, my first laser publication from Hughes was on an infrared xenon laser. It had high gain and the potential for operating in a straight-through amplifier mode. I spent about a year with that laser; a couple of publications evolved from it.

In late 1963, when I was still working on this infrared laser, a very interesting publication appeared from Earl Bell and Arnold Bloom at Spectra-Physics on the mercury ion laser. It was interesting because it put out more power than some of the things we had seen before, and it was visible. They reported red and green lines in a short publication, which just described the wavelengths and some of the conditions of operation.

I had the freedom in those days to jump on things that were interesting, and that was the kind of thing you just can't leave alone. So Bob Hodge, my technician, and I put together one of these mercury ion lasers to see what it was like and what the pumping mechanisms might be. Bell and Bloom's original publication reported operation with a mixture of helium and mercury. I had done some work before with the helium-xenon laser and found that the commonly accepted idea of what role helium played turned out not to be true. So I wanted to see if a similar problem existed here. Maybe helium wasn't really necessary.

So we built one of these helium-mercury lasers, got it operating in the lab, and then started playing around with it to see what its characteristics were. Is helium essential or not? Is it really charge exchange? Are these Penning collisions? What's the pumping mechanism?

One way to determine that is to take the helium out and substitute neon. I had a nearly complete station of gases at hand so we used neon as a buffer gas — that works. Actually, the laser will work very well with pure mercury. Anyway, after we had made a neon-mercury laser, we decided we would try argon as a buffer gas to see if we could make an argon-mercury laser. Later, we found out that works, too. But this particular day we couldn't make it work. We were using much too much argon. Nor did we have it well adjusted.

So, we pumped out the tube, flushed it, and put helium back in to make sure that our mirrors were still aligned. To our surprise, we found we had a new line going in what was ostensibly a helium-mercury laser. We now had a blue line at 4880 Angstroms in addition to the red and green lines from mercury. Well, that was very exciting and totally unexpected. It was a case of discovery rather than intentional construction.

L&A: That sounds like a combination of excitement and terror.

Bridges: I don't know if I can really convey the excitement. You see something unexpected, and furthermore you don't quite know how you produced it. Maybe it will go away, and you won't get it back again. To complicate things further, we didn't have spectroscopic-grade argon on the station. To do this experiment we wheeled up an old welding bottle with welding-quality argon and God only knows what possible contaminants. So we couldn't merely assume we were looking at an argon laser line.

We measured the wavelength to within a couple of Angstroms with a small spectrometer we had. Then it was hurriedly off to the library while the laser was still running. I fumbled around in the library, because I was unfamiliar with the spectroscopy of ionized argon and didn't know where to look. But after about an hour, I concluded that we were looking at ionized argon.

Well, with that confidence, we felt we could start tampering with the laser. So we pumped it out, flushed it with helium again, and still had the line, but weaker. Several flushes finally made the line go away. We were now back to having a helium-mercury laser. Then we let a little argon back in and reproduced the blue line.

We had already made pulsed time measurements on the mercury laser. As I recall, Bloom and Bell didn't mention that the mercury lines lased in the afterglow and didn't go with the current pulse. But in our laser, the argon line was very clearly going directly with the current pulse and the mercury lines were way in the afterglow. This wasn't especially encouraging, because the currents and voltages we were using in this particular pulsed laser were hideously high. But it seemed as though an argon laser might operate in a continuouswave mode, although only at incredible inefficiencies.

L&A: What happened to that first tube?

Bridges: The Smithsonian asked for the first ion laser not too long ago, but it had gone out in the trash the day after we discovered the argon line. We obviously were having trouble; we tried to get the mercury out of the tube so that we had just argon going. But once you get mercury in as a contaminant — and notice how quickly our thinking changed, considering that this had been a mercury laser just hours before — you can't freeze it out. We could almost make the mercury lines go away, but not completely.

We decided that the only way to do this right was to start with a clean tube. So this one was cut off the station, the glass blower was given a rush order, and by the next week we had a clean tube running with just argon. That was even more exciting because now, instead of just the 4880 line, we quickly had ten lines going in argon. That created a lot of activity.

L&A: Was this the point at which you published your findings?

Bridges: Now we were stuck — should we publish immediately or make careful measurements and work things out? We really wanted to nail down the wavelengths very carefully to identify these lines. All we had in the lab was a very small monochromator; you could measure wavelengths to within maybe 1 Angstrom or so. That isn't good enough to make a clear spectroscopic identification. But way down in the other end of the building resided a big two-meter Bausch & Lomb photographic spectrograph. We wanted to use that. But we couldn't move the laser, and we couldn't move the spectrograph.

We gathered a bunch of c-clamps and mag bases and put seven mirrors down the hallways, around corners, and through laboratories.

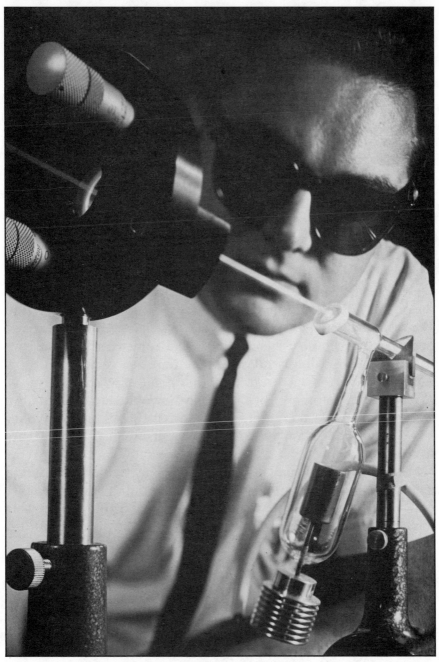

Bridges examines the intracavity radiation from a pulsed argon ion laser in 1964. This was "tube no. 2," constructed just one week after the original tube first lased on February 14, 1964.

We managed to move the laserbeam clear down to the other lab, probably three or four hundred feet away. While we were doing this, we were trying to keep it quiet. After all, we hadn't published any of this yet. When you have visitors constantly in a research laboratory, it's very difficult to explain what all those mirrors are doing in the hallways.

We usually operated at night. That was also a little safer, in case one of the mirrors drooped and we scattered laser light all around the halls. We ended up taking some very good plates and making measurements to within a few hundredths of an Angstrom that really nailed down the identifications. I am glad we did that because one of the lines had a rather illogical identification. There was another quite logical, rational line (that doesn't lase) that we could have identified it as. The precise measurement paid off.

We got this thing into print in *Applied Physics Letters* in April, 1964. As you may know, there were two other independent discoveries of the same argon ion laser. One was by Guy Convert at CSF in France and the other by Bill Bennett and his graduate students at Yale University. Convert and his associates published in *Compt. Rend.* and Bill Bennett published in *Applied Physics Letters*, but somewhat later than we did. So it was obviously a laser whose time had come.

L&A: Did you then go on to the other noble gases?

Bridges: Even before we got the publication written up and off, we started looking at other things. We had other gases on the station, so we went through krypton, xenon, neon, and some mixtures. Lines just sort of tumbled out all over the place.

Neon didn't work initially, and that's reasonable, because the analogous lines were all in the ultraviolet. We didn't have ultraviolet optics covering that range. We ordered some, but before we were able to get anything going in neon, Roy Paananen at Raytheon published his work with ultraviolet lines in neon.

Art Chester, who was a Howard Hughes doctoral fellow at Caltech in the physics department, spent a day a week at Malibu and was assigned to our group. When we started getting this real plethora of lines coming out at us, Art was practically a commuter between Malibu and the Caltech library, where he was photocopying old spectroscopic literature. Unfortunately, noble gas literature is *really* old. Those are the easiest discharges to make, so they were done first. The literature was mostly from the 1920s and 1930s, with more modern, very precise spectroscopy done on other materials. No one went back to redo the noble gases. As we improved our laser setup and went to higher current pulses, we were quickly producing lines that just didn't seem to be in the well-identified literature at all. Then it really became a puzzle to identify lines. Some of the lines we saw on xenon are still unidentified but are probably triply ionized xenon.

A few months later I was in Washington and went over to the National Bureau of Standards to get some help on these spectroscopic identifications. There I met Charlotte Moore Sitterly for the first time. She presided over the collecting and cataloging of spectral lines and is the author of the famous volumes entitled *Atomic Energy Levels.* She indicated she wasn't aware of any publications I hadn't looked at, but she suggested I might be interested in Curtis Humphreys' work. Of course, I had already seen his publications from the 1930s. I said, "Oh, are his notebooks still around here? Maybe there was something in his notebooks."

She said, "No, he took them with him."

"Oh, that's too bad," I replied.

She said, "Oh, he's still alive. Why don't you go see him? He is at the Naval Weapons Center in Corona, Calif."

Later, I called him and Curtis was kind enough to send me his ledger papers, hand written in ink, listing his laboratory observations. Happily, we found most of the funny xenon lines in that listing. Now, he hadn't identified them either. But at least it was an independent confirmation that somebody had seen the same lines 30 years earlier, so we weren't crazy.

Art Chester and I put together a publication summarizing all the spectroscopy that we did in this interval. We had a little note added in proof saying that we found this unpublished list of lines that suggested we were probably looking at xenon and not something else. Now, "not something else" wasn't merely an academic question. The tubes we put together used hot thermionic cathodes made of barium oxide. We encountered several other lines we had trouble identifying at first, but after the pattern fit into place, we identified them as ionized oxygen. Our discharge was partly disintegrating the cathode as it ran, contaminating the tube. We also found nitrogen and carbon ion laser lines from contamination and some other things we still haven't really identified.

L&A: How did continuouswave operation come about?

Bridges: A few weeks before the first paper was published, I went to a committee meeting, in New York I think, along with E.I. (Gene)

Gordon of Bell Labs. I had told him of the mercury laser work sometime previously. At this meeting I gave him a quick rundown of the argon work and preprints of the argon paper and a krypton-xenon paper that came out sometime later. I told him about the argon pulse following the current pulse, so that it might be a cw laser if it didn't chew up so much power.

About two weeks later, while Art and I were deeply involved in spectroscopy, Gene called me and told me, "We've got ours going cw." That surprised me. Gene had taken the leap of faith that it could be made to run cw. He took the approach of using a very tiny capillary discharge, about a millimeter in diameter, whereas our tubes were five or six millimeters in diameter. So his current density was up by a factor of 25 over ours for the same current. He also had a superb craftsman making mirrors for him at Bell Labs.

That resulted in a very interesting publication which a lot of people have questioned me on. Gene asked if I would be a co-author on their paper. I thought I ought to do something, since my entire contribution to that point had been to give him the preprints. So I offered to build a little different tube and use our bigger power supply to try to push on to higher power.

I started on that while Gene and Ed Labuda made the first draft of their work. They sent the draft to me and within a week I had a water-cooled tube going with something like 80 milliwatts output. I also got ours running cw with krypton and xenon.

That's how a joint publication between Hughes and Bell Labs came about. It was really just a matter of personal contact. I don't know how happy our supervisors were, but we did it anyway.

L&A: How long did you stick with ion lasers and the spectroscopy?
Bridges: I stuck with ion lasers for about six years; the spectroscopy was an on-and-off thing. You have to understand the environment of the times and the environment of Hughes. We had a cw ion laser with almost 100 milliwatts of output just as the first publication hit the streets in April, 1964. The Hughes systems people in Culver City got on us. They said they had looked at a system about a year earlier and decided that it just wasn't going to be practical with perhaps 50 milliwatts of helium-neon output. But now that we had about 100 milliwatts of green light, they asked if we could give them maybe a quarter of a watt, with some development, for this airborne system. Since we were looking at a laser whose power output went up with

the square of the current and didn't seem to stop, it was an easy extrapolation to a quarter of a watt. Too easy, as it turned out.

So, starting in May, 1964, when everybody else was jumping on spectroscopy, we started trying to build a demonstration tube that would give us a quarter of a watt cw and be packageable in an airborne system. The systems people, for their part, started putting the system together.

Through the summer and fall of 1964, we continued to do some spectroscopy evenings and weekends. But our days were mostly spent beating on the laser, doing very practical nitty-gritty development things. Success came fairly quickly. We had a quarter of a watt in a month or so and half a watt not too long thereafter. That was just as well, because the systems people kept coming back and asking for another factor of two. By late 1964, they wanted a two-watt airborne ion laser, so that's what we were shooting for.

The system, to use the *Aviation Week* description, was a scanning night-reconnaissance system. It had a spinning mirror that scanned one dimension on the ground, while the aircraft motion provided the other dimension on the ground. You ended up making a strip map of what's on the ground. Perkin-Elmer had pioneered this system and had one flying with a helium-neon laser in it.

Other applications came and went over the next several years, but our motivating force was a practical airborne ion laser. The first flight test was in 1966; by 1970 the lasers and the flight tests had become quite sophisticated. But by about 1972 the Air Force, for their own reasons, decided they weren't interested in that particular system. It was expensive, and it worked well. But it had some interesting ramifications on cooling in an airplane. Perhaps that was what made the Air Force a little less than enthusiastic about the system. But when that system interest ended, I think Hughes' interest in argon ion lasers ended, too.

L&A: Who were the key people in argon lasers at Hughes during this time?

Bridges: In the 1965 to 1970 time frame, the real players at Hughes were myself, Steve Halsted, Howard Friedrich, Peter Clark, and Neald Mercer. Steve came to us from Stanford University and later, about 1969, went to the Electron Dynamics Division as a department manager and took the argon ion laser into commercial production at Hughes. Steve brought what was probably the nicest low-price argon ion laser to production at that time, the old 3066H. But when that ran its course, Hughes got out of the commercial ion laser business as well.

Neald Mercer had been one of Bill Bennett's graduate students at Yale during the discovery phase. After Neald finished his thesis on ion lasers, he came to work for Hughes. Neald and I spent a bit more than a year working on an ultraviolet version of the argon laser, starting in about 1967. We were able to carry our spectroscopy into the ultraviolet and attain cw operation at new wavelengths. After about a year's work with tungsten disk-bore tubes, we were able to deliver a tube to Fort Monmouth NJ with a 2-watt output in the ultraviolet. Having that kind of laser available allowed us to push some of the lines that had only been pulsed before into cw mode and reach some higher ionization states.

L&A: Considering the state of ion laser development today, are you surprised by any of the commercialization improvements, such as the 25 milliwatt air-cooled versions available these days?

Bridges: No, I don't think there's anything really surprising in any of these lasers in terms of fundamental breakthroughs. They all represent very nice, careful, mature engineering. One thing that often surprises me is that some of the problems we faced in the early days are still around.

If I go back through my files, I will find a rather lengthy proposal we wrote about 1968 to study the window degradation problem. The government wasn't interested enough to fund it then, and commercial companies weren't really involved at that point. But the problems of window film formation and solarization are still around. Everybody has solved the window problem at least ten times over, only to have it come back in one or another form. It's one of those perpetual problems. I'm sort of surprised it hasn't been solved by now.

L&A: How about the emerging tube-refurbishment market, where a processor repairs and refills argon laser tubes?

Bridges: Well, we did it all the time. When tubes would fail for one reason or another, we would rebuild them ourselves over and over again as research vehicles. In fact, during our ultraviolet development work, we had one carcass with cathode and anode that had all kinds of different disk structures put in it. It was always a challenge to see how many times you could let a cathode down to air and still have the thing work.

Occasionally, I get a question from another faculty member here when they look at the price of a replacement tube. They say, "But it just looks like the windows are dirty. Can't we just pop them off and clean them? Why should I pay $12,000 for a new ultraviolet tube?"

Invariably, my advice is for them to bite the bullet and buy the new tube. If they only knew what was involved in reprocessing one of these big tubes, they would never undertake it on a one-shot basis. Even if they count graduate labor as free, it won't work out. I tell them, "You don't want to go into the ion laser research business. You are a chemist (or an aeronautical engineer). Just buy the tube." But in a commercial sense, it's probably a different matter entirely.

L&A: What did you go into after the ion laser effort wound down?

Bridges: I had become the department manager of what was then called the laser department at Hughes in 1969. If you will recall, there was something called a recession in progress then, so it wasn't a good time to break into management. After about a year and a half of layoffs and transfers in and out, I decided I was too young and tender to do this kind of thing. I really wanted to get back into the laboratory environment, so I resigned as department manager. Peter Clark, who had been my second in command and who had been managing most of the high-power laser activity, took over the department.

I was on the director's staff for a while after that. I found it was very difficult being on the director's staff. The director runs the whole place but doesn't own any lab space. So I had to mooch lab space from the departments.

After a while, I got involved in space communications. The real spark plug there was Frank Goodwin, who, by force of personality, kept laser communications alive through some pretty grim days. I developed an interest in the area when I was department manager. In fact, Goodwin, Don Forster, and I put together an analytical comparison of alternatives for laser space communications, in which we attempted to show the superiority of a coherent CO_2 system over an incoherent Nd:YAG system. Given that no one has put either system up yet, it is still an open question.

L&A: Wouldn't that require some significant tuning capacity on the part of the CO_2 laser?

Bridges: For a heterodyne system, you have to track out the Doppler shift from a moving satellite. That meant you had to have a tunable CO_2 laser with perhaps a gigahertz of tuning range. But a CO_2 laser isn't tunable to a gigahertz in its normal form. Part of my time on the director's staff was spent trying to work out some solution.

In mid 1971 Peter Smith at Bell Laboratories published a short paper on the helium-neon laser. The capillary bore of his laser was only about half a millimeter in diameter and long enough that the bore acted as a waveguide. Don Forster came back from a meeting and said that this might be the thing for CO_2 as well. Incidently, Marcatili at Bell Labs had proposed this structure in 1964; it's amazing that no one built one for seven years. (Actually, Steffen and Kneubuhl had unintentionally built a far-infrared waveguide laser in 1967 or so.)

In any case, I got a project going again with bootleg laboratory space and a borrowed technician to try to build a CO_2 waveguide laser. But I had to put it on the back burner for a while a few months later to assist a classified systems program that needed an injection of instant help.

By the time I got back to the waveguide laser in early 1972, Dick Abrams had come on board from Bell Labs. Peter Clark suggested that he collaborate with me on the waveguide laser. We spent some time on that, from which the first paper appeared in 1973. But I was also involved in other things, since I was on the director's staff. It looked like a good thing for Dick to do, so he took off with the waveguide work.

Then I was reassigned again, probably because I said something was ridiculous and impossible. The powers that be said, "Good, you're just the guy to work on it." That was the adaptive optics work that Hughes was trying to break into. At that time, Rockwell Autonetics had the only government contract, working with a ponderous CO_2 system. Tom O'Meara, our theoretician, had some ideas for a much simpler system. Frank Goodwin and Tom Nussmeier had been working on it, but they were reassigned. So making a working version of O'Meara's multidither prescription, as it was called, became my task for 1972.

Later, we got a very nice 18-element system going, using a visible-wavelength laser. That allowed us to make a movie of the system in operation, which is the only way to show off such a dynamic system.

By about mid 1973, the work had grown, and we had hired more staff. Jim Pearson joined us from Caltech and took over the lab work; I found myself managing again.

L&A: You have been here at Caltech for seven or eight years. How did that transition occur?

Bridges: In 1974, I had the opportunity to come to Caltech as a Sherman Fairchild Distinguished Scholar. So I did, with a bit of malice aforethought. I figured if I left Hughes for a year, they would have to find other people to manage these programs. Then when I came back I could become a worker again, because the managers would already be in place. It never occurred to me that they would find a manager and then have the manager split the scene about the time I was ready to come back.

I had a great deal of fun teaching a class on optics that year and even did some fiddling around with ion lasers again. I still wasn't able to identify those xenon lines.

I went back to Hughes with a dandy idea for laser isotope separation. I was even able to persuade Hughes management to let me do it. But the price extracted for that was to get involved in another area — the beginnings of the space-borne hydrogen maser clock. That was Harry Wang's project; he was the maser man and my contribution was to work on the atomic hydrogen source and the vacuum system. It was quite an engineering problem: How do you seal this device in a can and put it in a space craft to work for ten years, when it has a bad history of requiring a handful of Ph.D.'s to run it?

Between that and preparing for the isotope separation project, I spent a quick two years. I must confess that the isotope separation project never really got off the ground, because delivery on a tunable dye laser was delayed about 14 months. But in 1977 I had the opportunity to come to Caltech as a full-time professor. I'm not doing much in the way of laser work anymore. I'm spending most of my time on millimeterwave technology...particularly some novel dielectric waveguide components. You might call it a sort of integrated optics for the millimeterwave region.

But ion lasers are still fun. I like the engineering side of it as well as the scientific side, so I hope to continue on with ion lasers. I have done some consulting for Spectra-Physics, so I still have my hand in it. I may even set up one in the lab here for occupational therapy, to take another shot at those unidentified xenon lines.

L&A: If you had it to do over again, would you still get involved with lasers?

Bridges: You bet! Those were exciting times, the kind of thing that happens only once in a lifetime. I wouldn't have missed them for anything.

WILLIAM T. SILFVAST

Metal-Vapor Lasers

For sheer number of new lasers discovered, it's hard to match the record of William T. Silfvast. His doctoral research included demonstrating laser action for the first time in the vapors of nine elements. Much of his work has involved metal vapors, with the most tangible commercial result being the helium-cadmium laser. He also has demonstrated over 100 recombination lasers, laser action in laser-produced plasmas, and high-gain photoionization lasers pumped by soft x-rays. His most recent work is on short-wavelength lasers.

A native of Utah, he received two bachelor's degrees, in physics and in mathematics, from the University of Utah in 1961, and received a doctorate in physics from the school in 1965. During a year of postdoctoral research at Utah, he discovered the 441.6-nanometer blue HeCd laser.

After spending the following year at the Clarendon Laboratory of Oxford University under a NATO Postdoctoral Fellowship, he joined the technical staff at AT&T Bell Laboratories in Holmdel, N.J. in 1967. He has remained at Bell ever since, except for a 1982-1983 sabbatical year spent at Stanford University working on short-wavelength lasers under a Guggenheim Fellowship. He was made a Distinguished Member of the Technical Staff in 1983.

Jeff Hecht conducted this interview on September 28, 1984, in Silfvast's Bell Labs office.

L&A: How did you get involved with science?
Silfvast: From the earliest I can remember, I wanted to be a mechanical engineer. I always liked to tinker. But when I was studying engineering

at the University of Utah, I got pretty frustrated, because all I was doing was memorizing formulas. After two and a half years I quit and went to California with a couple of friends looking for a job. I wound up doing engineering work for Lockheed on the Polaris missile project, and after almost two years of that I realized that I had to go back to school and that I didn't want to be an engineer. I decided to go back to Utah and study physics. I enjoyed it so much that I decided to go on to graduate school, even though I hadn't planned to. In my senior year I took an optics course and fell in love with optics. It was just about that time that the laser was discovered, and I became very enthused about lasers as well. I started working with Grant Fowles, who was also getting interested in lasers. Before long, he and Russell Jensen discovered the first charge-transfer laser in iodine vapor.

L&A: What was your role in this?

Silfvast: Fowles thought it would be worth trying to make a laser out of bismuth to study hyperfine structure, since it is one of the heaviest of the odd elements and consequently has a large hyperfine splitting. It also has a reasonable vapor pressure. I accepted the challenge and scrounged some zone refining furnaces from the ceramic engineering department as well as other miscellaneous parts and put together a laser discharge system.

There had been a couple of metal vapor lasers, in discharge-excited mercury and optically pumped cesium. I was not aware of the cesium work at the time, and the mercury laser didn't require heating to vaporize the metal. The mercury work, however, no doubt inspired us. We didn't have much financial support, but Fowles was ingenious at putting together inexpensive apparatus, and I adapted quickly to that. We made a quartz tube that we could just attach to simple electrodes and used microscope cover slips mounted with vacuum grease as Brewster angle windows. I think that system was unique then, because I could change from one metal to another in a couple hours and try a lot of experiments in a short time.

L&A: That must have been one of the first systematic searches for new lasers.

Silfvast: It turned out that it was, but we didn't start that way because we were focusing on bismuth. We didn't know how long the metal would last and thought we might have to change tubes very often. Also our approach was inexpensive, if you had a good glass blower.

That tube design turned out to be a key element in our success.

William T. Silfvast

Later we found out that some people in Gene Gordon's group at Bell Labs were thinking about metal vapor lasers. They apparently were slowed down by building elaborate high-vacuum discharge systems, and we got our results first. Also, we didn't have preconceived notions of how pure a laser tube should be—it had to be very pure in the types of lasers discovered before then—and in some ways our naivete was a help.

L&A: What happened with bismuth?

Silfvast: I spent a good three or four months looking for lasers at various wavelengths, mostly in the visible, but never saw a laser at all and was somewhat discouraged. Then one day in early 1965 I said, "Hey, we're not getting anywhere," and thought about mercury. Looking at the periodic table, I saw that cadmium and zinc had similar electronic configurations. So I went to the chemistry department stock room and obtained some purified zinc and cadmium.

I put zinc in the tube first, and the very first time I turned it on I got this turquoise, blue-green transition at 492.4 nm to lase. It was on a Saturday. I was just ecstatic and went running around the campus looking for Fowles. I found him in a committee meeting, which he immediately left and came running down to the lab to see the laser.

We studied that system for a few days, and then we put cadmium into the discharge tube and made many of its transitions lase, although not the blue one yet. The transitions we got to lase were those that seem to work by charge exchange. We decided that many of the metals were the most interesting candidates for lasers and started trying every metal we could find that we could vaporize in our quartz discharge tube, which went up to 1200°C. Besides cadmium and zinc, we made lasers in gallium, germanium, tin, indium, and so on, but the next really exciting one was lead.

A lot of the early gas lasers were discovered somewhat accidentally by researchers looking for another transition. Gas laser excitation mechanisms were too complex to model accurately at that time, though we do a lot better now. We really were trying to make lead ion lasers. In fact, we had made them. But one day, when we had blue mirrors on, I saw what was obviously a red laser beam. It was exciting. I've discovered my share of new lasers, and every one was very exciting because it was something new that you were seeing for the first time. In this case, I said at first that this can't happen—the mirrors must have had losses of 70% to 80% in the red, yet I saw the laserbeam.

It turned out to be the high-gain self-terminating neutral-atom lead laser that led to the whole series including manganese and copper.

Bill Bennett heard about our work and flew out to see it. We were impressed that he came out; he was one of our heroes. We didn't know that at the time he was consulting with TRG where they were working on Gordon Gould's idea of a collision laser. They had been running a continuouswave discharge in manganese, and we were later told that after Bennett saw our work he told them to start pulsing their discharge. Later, while we were making manganese lasers, all of a sudden Fowles got a letter on TRG's work in manganese to review for *Applied Physics Letters*. They had scooped us, though I'm sure in the long run it was good to have Bill Bennett visit us.

L&A: Were other laser researchers surprised at this little group way out in Utah?

Silfvast: Fowles gave a paper at the Electron Devices Meeting in Urbana IL. Besides zinc, cadmium, and lead, we also had made phosphorous and sulfur lase. We scooped two Bell Labs groups! Colin Webb, Ed Labdula, and Dick Miller approached us to say that they were thinking about making metal vapor lasers. And later Peter Cheo told us he had planned to give a postdeadline paper on phosphorous and sulfur lasers at another meeting in a month or two. I can't tell you how thrilling it was for me because I thought of Bell Labs people as the ultimate leaders in research.

L&A: What did you do after that?

Silfvast: I did my thesis on the lead laser. I predicted some incredibly high gains, 6 to 10 decibels per centimeter, much more than anybody else had reported back then and higher than we knew how to measure. Then I stayed on at Utah for another year as a postdoc. One of the things I did that year was study some peculiarities I had seen in the laser characteristics of some of the metals. We had seen bright blue emission, not laser emission, from a couple of zinc and cadmium transitions. I went back and studied them. In 1966 I first observed the 441.6-nm blue cadmium laser and found that it worked only with helium. We hadn't made it lase the first time around because it required very low excitation.

L&A: Had you been working with nominally pure metal vapor before?

Silfvast: No, always with a buffer gas to get the discharge out of the heated region to the electrodes. We hadn't realized that helium was essential for some transitions. We were going through so many lasers that it would have taken at least 10 years to work out the details of the excitation mechanisms.

I applied for a NATO Postdoctoral Fellowship to work with John Sanders' group at Oxford University—he was doing some laser work, and I liked the idea of going to Oxford. I got the fellowship, which was the first thing I had ever won in my life, and it was very exciting. One of the things we did at Oxford was to develop a technique for measuring very high gains that showed the lead laser gain was 6 to 7 dB/cm, agreeing with the calculations I had made earlier.

Meanwhile, Bruce Hopkins, a graduate student at Utah, continued studying the HeCd laser. He saw some quasi-continuouswave emission in the blue, at the peak of the ac excitation cycle. When I came back and visited, Fowles showed me the results, and they looked promising for a continuouswave laser.

L&A: You did that work here at Bell Labs. How did you come here?

Silfvast: I interviewed with Bell Labs while I was at Oxford and was very excited when they made me an offer. I was very surprised at the salary, several thousand dollars more than I would have accepted as a minimum to work at Bell. At first I came only for two years. I wanted to teach, and I think Bell had not seen enough of me to risk offering a permanent post. But I liked it so much here that, when they offered a permanent post, I took it. I didn't have strong training in optics as a student, since PhD exams were offered only in solidstate and nuclear physics. It took me a good ten years to get over my feelings of inferiority at Bell, even though things went very well for me here.

L&A: Did you start right in on metal vapor lasers?

Silfvast: Not exactly. First, Kumar Patel suggested I work on carbon-dioxide lasers, something of more interest at that time than metal vapor lasers. So I started coming up with some ideas there, but meanwhile I talked more with Patel and P.K. Tien, who was my department manager. I convinced them that there were some interesting things in metal vapors. Also, Patel was working on self-induced transparency, which was a hot thing back then, and we thought that the lead laser would be useful for that. I started dividing my time between the lead and cadmium lasers.

I came here in August, 1967, starting with a completely empty lab, and by November or December I had a HeCd laser working continuouswave for the first time. By the following June, I was able to give a postdeadline paper at the International Quantum Electronics

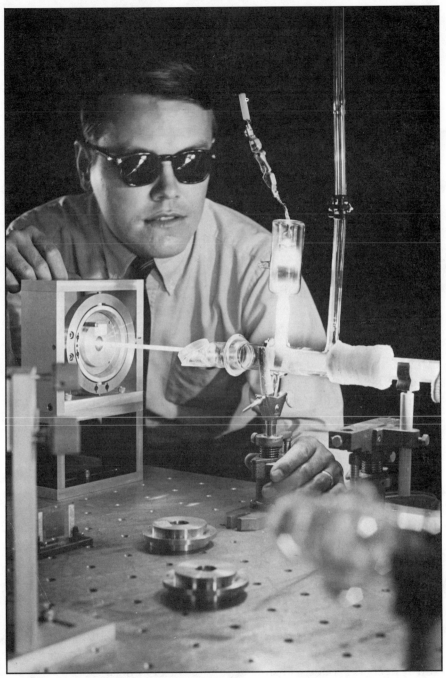

Silfvast demonstrates one of the first continuouswave HeCd lasers.

Conference in Miami reporting continuouswave, fairly efficient operation of the 441.6-nm HeCd laser.

L&A: What was the key to cw operation?

Silfvast: Learning how to run a steady dc discharge in a metal vapor such as cadmium, running a low enough excitation current, and getting the proper vapor conditions. At Utah we used ac neon sign transformers that we had scrounged. They already had built-in ballasts and were much easier to operate in controlling the vapor pressure when we were just looking for lasers. Running a dc discharge is harder because you have to worry about contaminants in the discharge and cataphoresis effects. The conditions we needed were quite different from those in the argon laser which had already been discovered by Bill Bridges.

Tom Sosnowsky, here, discovered the cataphoresis effect, and I think John Goldsborough also did at Spectra-Physics at the same time. We had been producing the metal vapor by putting the metal in little cups or dimples in the discharge bore. Then Sosnowsky found that cataphoresis in cadmium is high enough that a single source of metal at one end of the tube can produce a fairly uniform discharge over the entire length of the tube. That was a significant step in the development of a commercial product because it made the tube design simpler.

L&A: Who discovered the 325.0-nm ultraviolet line of HeCd?

Silfvast: John Goldsborough and I did that independently, but I think his publication was first. The discovery of cataphoresis helped make a more uniform gain region which helped overcome both the mirror problems at 325.0 nm and the reduced gain on that transition.

L&A: How did the HeCd laser become a commercial product?

Silfvast: I think Goldsborough built the first model at Spectra-Physics in 1969 or 1970. I inherited one of those old monsters from somewhere, and it's still lasing. It was a couple of meters long, very expensive at the time, and it was easy to see that it would never be practical. RCA came out with a similar laser not long afterwards.

About that time, a fellow from a development area here came to me with a new idea for a remote blackboard. The idea was to transmit in real time only the changes in the chalk motion on a blackboard during a lecture. This would take only the bandwidth of a single telephone line. If you were scanning the entire blackboard for changes, you'd need the bandwidth of a television channel.

They found that the HeCd's ultraviolet line could write on a DuPont product called DyLux and wanted to use laser writing and projection of the image for a display. They came to me looking for a small, inexpensive ultraviolet HeCd laser. I had some ideas about what we later called the segmented-bore HeCd laser. If they worked I could build a HeCd laser that would fit into a suitcase, so I started on the project.

We put little cadmium washers between glass discharge tube segments and demonstrated 1,000 hours of operation in this laser, which was self-heated by the discharge. It gave 15 milliwatts in the blue, as much power as the RCA model that was three or four times as long, and it also produced 2 milliwatts in the ultraviolet. No one had ever made a laser that short work in the ultraviolet before. And it used the natural mixture of cadmium isotopes instead of the costly single isotope.

I was doing this all on my own, and with that success I was sort of committed. At one point I mentioned to management that these development people wanted a laser, and they said, "We don't want to be a laser supplier, that's not our purpose here." But I kept working on the project. That's the freedom that we have here at Bell Labs.

After I had demonstrated the laser, the remote blackboard people got very excited. The laser was good enough to let us think about a commercial product. So I sent descriptions of my work to five companies, looking for two of them to build three or four development models each for the remote blackboard system. Some companies said it couldn't be done, but by the time we got out to Coherent Radiation (Coherent Inc.), they already had one working. They had some good ideas, so our development department gave them a contract and also gave one to Hughes, which had other good ideas.

Jim Hobart at Coherent liked the laser and put Mark Dowley on the project. Mark worked on it for about two months and decided it was such a good idea that he would start his own company. So he left Coherent and started Liconix.

L&A: Whatever happened to the remote blackboard?

Silfvast: The ultraviolet laser was not far enough along in development, and they found out they could use a television display instead—it's a commercial product now. Tony Berg, who was then at Bell Labs, also built a very competitive scanner and microfilm printer that used the blue HeCd laser. However, even though it was a good system, it never made it to the marketplace—it wasn't the kind of thing that AT&T was geared to sell in those days.

L&A: Has anything held back commercial development of HeCd?
Silfvast: There were some problems in the early days with the company that advertised an inexpensive HeCd laser before their laser was proven. They sold many lasers at a very low price. When their laser suffered problems and was withdrawn, this gave the laser a bad name for awhile. But the real problem has been a lack of investment in the technology to bring prices down. The HeCd discharge tube can be made almost as inexpensively as a helium-neon laser. The power supplies do have to be more expensive than HeNe, and because you have to heat the discharge, it is difficult to use internal mirrors. Both Liconix and Omnichrome have some very good lasers for sale now with 5,000 hours or more of operating life, but the real low-cost laser has not yet happened.

L&A: Why has HeCd been the most successful metal-vapor laser?
Silfvast: Because it is simple and has the right wavelengths. Cadmium happens to have the right electronic configuration. By just removing an inner shell *d* electron from the ground-state configuration of cadmium, Penning ionization from helium metastables or direct electron or photon ionization of cadmium puts you in the upper laser level of the ion. This is in contrast to the normal situation where removing a single electron just puts you in the ion's ground state, which of course can't be inverted with respect to other ion states. Similar levels exist in zinc, but the transitions are in the red and yellow, which are not as interesting, and zinc needs a higher operating temperature.

L&A: Whatever happened to helium-selenium? You had that on the cover of the February 1973 *Scientific American*, and there were some commercial models in the mid-1970s.
Silfvast: I think HeSe can simultaneously put out more wavelengths than any other laser. We've been able to lase about 25 lines from the blue to the red at the same time. When Marv Klein and I first discovered it, there was no continuouswave dye laser, and we thought it would be a big commercial success. RCA and Liconix offered commercial models, but they had to pull them off the market because selenium vapor pressure is much harder to control than cadmium— it tended to wander around and end up in the wrong places. I believe those problems could be solved, but there hasn't been a strong application need to justify that investment.

The cover photo, which has become relatively well known, was an interesting story. Fritz Goro, the *Life* photographer, got the credit, but Leo Szeto and I actually devised the way to take the picture, which shows the many laser lines spread out in multiple orders from a diffraction grating. Smoke didn't work, so we finally painted the beams onto the film by moving a large white card back and forth.

L&A: What are you doing now?

Silfvast: I moved from the metal vapor lasers to recombination lasers, where excitation comes from recombination of a free electron with a positive ion-excitation from above. Obert Wood had made one of the first TEA carbon-dioxide lasers, which produced bright plasmas by focusing the pulses in the air. I thought we ought to be able to make a laser in such a laser-produced plasma. Eventually, in 1976, we succeeded in making the first laser-produced plasma laser, using recombination in rare gases.

Later Obert and I developed probably the simplest laser ever discovered, the SPER laser, for Segmented Plasma Excitation and Recombination. It consists of a series of electrodes—as few as two—that you can apply with double-sided tape, a couple of mirrors, and a low-pressure gas in the cavity. With proper electrode spacing, you can produce a metal vapor spark in the electrode gap region rather than a gas discharge, using either high voltage or—we found out later—low voltage. The high-density metal vapor spark plasma expands, and laser action in the recombination spectrum occurs in the periphery.

We've discovered over 100 new laser lines this way. It works with a lot of metals that we couldn't easily vaporize in other laser designs. Some designs are very easy to make. We call them BIC lasers because we think they could be sold for a couple of dollars if you made them in large quantities. They produce peak powers of a few watts in pulses 10 to 100 microseconds long, and we've had some ideas for applications. The wavelengths range from just under 300.0 nm out to 5 micrometers. Most are in the neutral species in the infrared. The visible ones are mostly in the single ion, and the ultraviolet ones mostly in doubly-charged ions, because energy levels get farther apart with higher ionization levels. We had hoped to go into the vacuum ultraviolet, but the SPER lasers can't produce significant excitation beyond the double ion. So we set that work aside, although it has interesting commercial applications, because our real goal was short-wavelength lasers.

L&A: Have you found a more promising approach?

Silfvast: Obert and I did some more work with laser-produced plasmas. Then in 1982 I received a Guggenheim Fellowship and spent a year with Steve Harris at Stanford. It was a very stimulating time, during which I got the idea of trying inner-shell photoionization using soft x-rays to directly pump cadmium. The equipment at Stanford was tied up with other experiments, but I did the modeling out there, and we set up the experiment back here. It was one of those unusual times where I had the chance to do all the theory first and consequently knew the exact experimental conditions to set up. It worked on the very first shot.

We focused a high-power laser onto a target in a heat pipe containing cadmium. That produced a plasma that emits x-rays, which pump the cadmium vapor to produce the particular *d*-electron state of the 441.6-nm blue transition, giving very strong inversions. Mike Diguey had predicted that these inner-shell transitions could be used for an x-ray laser back in 1967, but no one had ever made a laser this way until we did in 1983.

Since then, we've been trying to take this energy and transfer it to other states. The exact location of many of these states is not yet known, and we are involved in some new spectroscopy of autoionizing states. We are very excited about using this inner-shell process for short-wavelength lasers. I don't know of any other process that can concentrate the broad spectrum of soft x-ray pump photons to produce a single state, which is what you need to make efficient short-wavelength lasers. Everything else is working against you. Lifetimes are shorter, the gain cross-section is going down with wavelength, and therefore the densities and pump powers have to be considerably higher just to get the population in those states. We have some new results now but are still struggling with the diagnostics to see if there is gain in the vacuum ultraviolet. If we've got a laser, it will be the simplest vacuum-ultraviolet laser yet.

L&A: If you had it to do over again, would you do the same type of work?

Silfvast: It is so tremendously exciting to make a new laser that I can't ever imagine not looking forward to coming to work every day, even in slow times. I've probably made more new types of lasers than anybody else, if not in numbers of wavelengths at least in different species. Many of those new lasers have opened up new areas of research and development, and that's the kind of excitement that keeps me coming back for more!

JAMES J. EWING

Excimer Lasers

James J. Ewing received a doctorate in physical chemistry from the University of Chicago in 1969. After serving on the chemistry faculty at the University of Illinois in Urbana and the University of Delaware in Newark, he joined the technical staff of the Avco Everett Research Laboratory in 1972.

At Avco Everett he and Charles A. Brau were the first to demonstrate three key excimer lasers: krypton fluoride, xenon fluoride, and xenon chloride. He also worked on laser enrichment of uranium isotopes before moving in 1976 to the Lawrence Livermore National Laboratory, where he was program manager for the "Rapier" advanced excimer fusion laser technology testbed.

In 1979 he was named director of the laser technology group at Mathematical Sciences Northwest in Bellevue, Wash., now Spectra Technology Inc. He now is vice president for laser programs, responsible for managing and directing Spectra Technology's laser research and development business. That work includes developing free-electron lasers, gas-discharge lasers for laser radar, high-power and tunable solidstate lasers, electronbeam pumped and high repetition-rate excimer lasers, and short-wavelength drivers for intertial-confinement fusion.

C. Breck Hitz conducted this interview on December 13, 1985, in Ewing's office at Spectra Technology, Inc.

L&A: What motivated you to look for laser action in excimer molecules in the first place?
Ewing: Beginning around 1972, when I first joined AVCO and started

working with Charlie Brau, there was a big push to find efficient visible lasers. There had been success at making efficient 5- and 10-micrometer lasers, carbon monoxide lasers were looking like they could be 40% or 50% efficient, on paper anyway, and carbon dioxide was scalable to high power at over 10% efficiency. And there was work on hydrogen fluoride going on, some of it at AVCO but most of it elsewhere. It looked like one could make good lasers in the infrared, but the question was, "Why can't you make them in the visible?" And excimers were one of several candidates for good visible lasers.

L&A: Is there any one person that you can put your finger on as being the first to conceive the idea of an excimer laser?

Ewing: Well, you have to understand, there are two kinds of excimer lasers. People were thinking about pure rare-gas excimers for a long time, since even before I got into it.

L&A: You mean rare-gas dimers, like Xe_2 and Kr_2?

Ewing: Right. I'd have to look up who was the first person to come up with that idea. Jim Keck at AVCO had suggested using a dissociative state as a lower laser level many, many moons ago, and he was thinking about hydrogen. Houtermans had the concept of using mercury molecules as a laser medium. Around 1971 or so there was a variety of people who proposed different things in excimers. Charlie Rhodes, then at Livermore, was one of the most prominent, and he really got the first of those experiments on the air, working on pure rare gas excimers: xenon, krypton, and argon. There was rare gas dimer work at Sandia (Gerardo and Johnson), Northrop (with Bhaumik, Ault, and others) and work in the Soviet Union around the 1972 and 1973 timeframe.

Within AVCO Charlie Brau was the first to champion the original suggestion which came from Houtermans. We embarked on mercury-based excimers, but I don't think any one person could be credited as being first to say, "Excimer lasers, that is something to go after."

L&A: What about the rare-gas halide excimer? Can you pinpoint where that idea came from?

Ewing: First off, just a peculiar comment on English: "Excimer" is a word from organic molecular physics. It's short for "excited dimers," and "dimers" means two of the same thing. It's definitely a misnomer as we use the word in laser physics to refer to things like KrF. There is a right name for these molecules: "exciplex." But it doesn't rhyme with laser, so if you call it an "exciplex laser," that's

James J. Ewing

a big mouthful. "Excimer laser" sounds better. My own belief is that people just call them that because it sounds better.

But my involvement in the rare-gas halide work came about because of a meeting that I didn't attend. There were two university labs who were working on it ahead of us at AVCO. One was Don Setser, a professor at Kansas State. He had done a lot of work on energy transfer from argon's metastable excited states. He was doing kinetics in low-pressure flowing afterglows. He presented a paper at the 1974 chemical laser meeting in St. Louis. I wasn't there. But Hao-Lin Chen, Charlie Brau, and Bob Center from AVCO were there. They heard his talk, got excited, and came back and said, "Setser gave a paper on how he reacted xenon metastables with halogens and got this spectra which had to be produced by a molecule of xenon and chlorine."

But the real interesting question to me was, why on earth was the spectrum of xenon excimers at 170 nanometers while xenon chloride was at 300 nm, if they're the same kind of molecule? That was the key question, and that was a question that was posed to me by Charlie Brau, I believe on the next Monday.

So that's what originally got me interested in rare-gas halides. There was this basic property of excimers, we'd seen it in mercury excimers, xenon excimers, and things like that. The resonance level is at about 10 electronvolts and the emission is at 8 eV. There was maybe 1 eV of binding energy in the excited state, everyone could understand and believe that. And there was 1 eV of repulsion in the ground state. Ten eV minus two puts the emission at 8 eV, right where it belongs. So why was XeCl emitting around 5 eV? Well, I remember sitting down and thinking that the excited states of the rare gases are very similar to alkali atoms. They have very low ionization potentials. So one of the things that I realized early on was that these new molecules had a nature that was kind of like that of an alkali halide. And I put together a very crude model and discovered that, based on what happens in an alkali halide, it's not surprising at all that XeCl emits around 300 nm.

Next we said, "Let's try an experiment. Let's put some iodine and some xenon in front of our e-beam, zap it, and see where the light comes out." (Xenon iodide, by the way, has never lased, but it was the first thing we tried to make fluoresce.) According to my crude model, the light was supposed to come out somewhere around 250 nm. So we zapped it, and a tremendous amount of light came out.

There were two things we noticed. One was my wavelength prediction was within 20% of where it should have been. So we said, "Maybe there's some veracity to the model." Second, the emission was brighter than anything we'd been working on for the last three years. We said, "What have we been doing wrong all this time?" It was sort of an exciting thing. I remember sitting down and saying, "Well, if this is where it is, what really do those potential curves look like?" I came up with a different model that went into our first paper, which was in *Physical Review*. That was a classic story, in and of itself.

L&A: You had problems with journal referees?

Ewing: We sent this paper describing the new model to *Physical Review Letters*, and the referee sent it back saying, "Well, it's nice, but all it is is spectroscopy; you looked at some spectra, so what?" This was a classic case of missing the significance of a paper. So we said, "Send it to another referee." Another referee said, "Well, it's an interesting thought, but maybe you ought to look at a few more molecules to see if that theory is really correct." By this time we said, "Just put it in *Physical Review*." We didn't care if it came out in *Physical Review Letters*.

But it kind of galled me at the time. I remember, it was April Fools Day of 1975 when Charlie Brau put this made-up letter from *Physical Review Letters* on my desk. It said something like, "Well, we've sent your paper out to this reviewer and to that reviewer, and they think you ought to go back and remeasure everything," and so forth and so on. I looked at that; it made me so irate! But he was just pulling my leg.

L&A: So it was Setser's observation of fluorescence from XeCl that eventually led to your work with rare-gas halides?

Ewing: That's true. Independently Golde and Thrush in England were looking at argon excited state reactions with chlorine at low pressure. But one of the things that was curious was that the spectrum that we observed was entirely different from what the low pressure work showed, and that had to do with what pressure we were at. Setser did things at very low pressures, and his bandwidth was huge. When you excite the rare-gas halides at high pressures, the emission bandwidth is considerably narrower.

So, if you calculated the gain coefficient based on Setser's data, you might conclude that these molecules aren't very interesting for making lasers; the spectrum is so broad it would never lase anyway.

But at AVCO, we looked at Setser's data and asked, "Why is the emission at that wavelength in the first place?" And that was the right question to ask because it led to an understanding of rare-gas halide molecules, and that understanding led to the laser.

L&A: When was all this going on?

Ewing: Well, it was in '74 when we started working on this stuff. September or October of '74, and our first paper was ready to be submitted in December of '74. It wasn't until the next June that we actually got one to lase.

L&A: What was the first rare-gas halide laser?

Ewing: That was xenon bromide. I guess it was around March of '75, when there was a briefing at Naval Research Labs for the 50 or 60 people working on these projects. There were Phelps and Gallagher from JILA, Charlie Rhodes and the Northrop group working on Ar/N_2 plus all the people from Naval Research Lab, and others. We gave a presentation outlining where all of the excited states of these molecules should be, and where all the emission bands should be, and so forth. We showed the high-pressure spectra of all the xenon halides. Afterwards, Stu Searles of NRL pulled me aside and showed me their work. They started it about the same time, having been excited by Setser's talk.

What they were working on was XeBr and what they were up to was really different. We had a really good lock on the spectroscopy and then, ultimately, the kinetics. I think Stu was working at lower pressures initially. I don't think he'd tried to jack up the pressures yet. I don't know, something was different. We compared notes on some of the things and went away from that meeting. About a month or so later, Stu Searles pulled out all the stops on his e-beam gun and got XeBr to lase. It was a curiosity that XeBr was the first of those molecules that lased. It has never really seen much use anywhere because it's terribly inefficient.

L&A: XeBr lased in the spring of 1975?

Ewing: Stu Searles did his work around May 1 of '75. He and Hart beat us—beat us with the wrong molecule—but that's okay. So, Stu Searles and Hart did XeBr; we heard that and said, "Now why is it that their e-beam is pumping it, and we can't get xenon fluoride to go?"

And basically the challenge came down from above and from amongst ourselves: "Why is it that NRL's e-beam can do that and our e-beam can't?" We went in and changed our anode/cathode spacing and ripped out half of the foil supports in our hibachi, just threw that all to the wind. We finally had the nerve to turn up all the knobs and work at pressures where before we would have thought we were risking the foil.

It was about May 15th when XeF first went. I could go look at my champagne bottles at home, because I saved them. XeF was the first of ours to lase.

We had previously developed a certain conservatism while we were working with high temperature mercury, because anytime a foil would break, your whole diode would get filled with mercury and it was a mess. You were down for a month almost. But working with room temperature gases, it was much nicer. We put xenon and fluorine in and made those changes, and we got XeF to lase. That was quite a hoot! A week later we did XeCl. Then a week after that we got in our optics to do an old favorite of ours, the nitrogen-oxide gamma bands. We had optics that were coated for making the gamma bands lase. We had this mixture concocted that was argon, nitrogen, and NO, and the NO was supposed to lase. We spent a week working on it, but it didn't lase.

So then we tried krypton fluoride; my model said it was supposed to be at about the same wavelength. We put some krypton and fluorine in and looked at the fluorescence. KrF was by far the brightest thing we'd ever seen. We had been scared of using fluorine at the time. Even though others at AVCO were making chemical lasers with, by comparison, huge amounts of fluorine in them, it took us a while to realize that the amount of fluorine we had was miniscule.

Anyway, we tried KrF, and it fluoresced like mad. Then we put the mirrors on it, and unfortunately the mirrors didn't have real high reflectivity at 249 nm. We had about 20% or 30% output coupling. Even so, the thing just lased like mad. If we had had high reflectivity mirrors on it, it would have just fried them.

L&A: How good were UV optics back in those days?

Ewing: Not very good. They were not very common. That was always a problem. Even windows—we didn't even know how to treat windows. It's important in quartz to make sure you get all the water off, because if you put a little fluorine in there, it etches. We were seeing the effects of photochemical etching.

So we got KrF to go; that was June 5 of '75, I believe. We had several new lasers all in a row there, and that was about the time of the 1975 CLEO meeting. I remember Stu Searles gave his post-deadline talk on XeBr. Right after that I gave a talk on what we had done: KrF, XeF, and XeCl. It was exciting, and I think that accomplishment excited a lot of people. It wasn't just one off-the-wall laser, it was a whole family with promise. We had a good first cut on the kinetics. Charlie Brau realized that electron attachment followed by ion-ion recombination was fast and a dominant pathway.

That is when others started really getting on the bandwagon. Shortly after our work was presented at CLEO, Bhaumik, Ault, and Bradford at Northrop lased XeF using nitrogen trifluoride rather than F_2 as a donor.

The curious thing is that our first excimer-laser publication was about KrF and XeCl, which were really the third and fourth rare-gas halides to lase.

There was another one that lased at that time. There are two kinds of bands in rare-gas halide excimers. There are the ones that go to the main level, the really bright ones that are very sharp because the lower level is kind of flat. Then there are other bands that go to a different final state, bands which are very broad. We always had the idea we'd make the broad ones go.

I had done some calculations and some ruminations on that. Now XeI had the brightest broad bands of all of them in its emission spectrum. So we placed our bets on XeI. We concocted a mix of argon, xenon and I_2 or hydrogen iodide, I forget. We put it in, put on the same optics that we had used for XeF, and zapped it. Jim Dodge, our technician, got it to lase on the first shot! But the light came out at a wavelength that didn't really look like where it should be. So we all celebrated and had our by-now-traditional bottle of champagne. Then we asked, "What is it that's really lasing here?"

The lasing spectra didn't have the broad structure we expected from XeBr. It had all these fluctuations in it. "What is it?" we asked. "Well, let's take out the xenon." We took out the xenon...

L&A: And it still worked?

Ewing: It still worked. Swell... but without the xenon in it, it wasn't XeI lasing. It shouldn't be argon iodide either, because I had done some calculations that said ArI wouldn't have a bound excited state. So the only thing it could plausibly be was something related to I_2. We deduced that it was a state in I_2 that was probably formed by ion/ion recombination of an I^+ and I^- that somehow got made. So we had that I_2 laser in there also. A fourth bottle of champagne and a fourth paper. Subsequently Br_2 and F_2 were lased by various workers on similar transitions.

L&A: You've mentioned some of the people who were involved in early excimer work. Who else contributed?

Ewing: Before the rare-gas halide work, SRI (Don Lorents, Dave Huestis, Don Eckstrom, Bob Hill, etc.) had done some very nice kinetics on $ArlN_2$ and Hg_2 and HgXe. IBM and Los Alamos also worked on rare-gas dimers. Interest in rare-gas halides blossomed in the summer of 1975. At that time Northrop (Earl Ault, et al) and Charlie Brau and I at AVCO were pushing as hard as possible to understand and increase the intrinsic laser efficiency. Intrinsic KrF efficiencies of over 10% were reported.

Joe Mangano and Jonah Jacob were the first to demonstrate the discharge pumping of KrF in an e-beam-sustained discharge. Joe and Jonah worked hard on understanding these discharges. Ralph Burnham and Nick Djeu were at NRL along with Searles and Hart. Burnham apparently was the one who said, "Why don't we modify an old Tachisto CO_2 laser, give it a faster circuit, and see if we can make it into an excimer laser?" I think he and Nick got XeF to go first. That was a very, very significant thing. It got excimer lasers to the point where they were in principle no longer a laboratory curiosity, where they could be avalanche-discharge pumped. Brunham and Djeu did that in the summer of ick Airey was their supervisor at the time. Much of this work was first reported at Woods Hole in September of 1975. That meeting was a watershed in spurring further work.

L&A: What about the Sandia people; what was their contribution?

Ewing: They got involved in the summer of '75 after CLEO. They first talked about their effort at the Woods Hole meeting. They were the first to scale a number of these devices to a large number of joules. In fact, the first time argon fluoride was made to lase, sometime during the winter of '75-'76, they whacked out over 100 J from it. It was the first time the ArF spectrum was even recorded. To me that always was a mind boggler. But they had this huge e-beam— 3,000 J at a megavolt or something like that. They really put it to those molecules. That's what Kay Hayes and Gary Tisone were working on. But those people at Sandia didn't just come out of the woodwork, you know. They'd been e-beam pumping lasers for a long time. Wayne Johnson and Jim Gerardo were the big honchos in that effort. Those guys had done a very nice job on xenon excimers. But there were a lot of people involved, and I can't possibly recite them all. In the xenon VUV excimer work, Charlie Rhodes, Paul Hoff, and Buddy Swingle at Livermore come to mind. John Murray and Howard Powell were also at Livermore, working on rare-gas oxides.

L&A: To what extent were the Soviets involved in the development of excimer lasers?

Ewing: I think Danilychev had worked on e-beam pumped xenon, but they were really behind the US work on rare-gas halides. Even a year and a half after we had come out with the first rare-gas halide reports, I know Bill Krupke visited Russia and found there wasn't anyone there doing anything on it. They seemed to be about a year or so behind. In fact, the beauty of the Russian work was that they were behind and didn't read a lot of our papers, so they didn't have some of the prejudices that we had. One of our prejudices that came from our early work was that XeCl was not efficient. We thought XeCl was so because we were using Cl_2, we were using it very rich in halogen donor. The Russians didn't know about this and simply ignored it. They went out and made XeCl work better in a discharge laser than anyone in the United States had been able to. So the word came back to us, "Maybe we ought to look at XeCl again."

L&A: Were there any Europeans involved in the early work?

Ewing: The first publications on gaseous rare-gas halide emission was by Golde and Thrush in England, but they were doing some physical chemistry experiments, and the possibility of lasing wasn't considered. They saw argon chloride emission at 172 nm, and they did their work just slightly before Setser. So if you look at who is the first person to mix some electrically excited xenon or argon or whatever with a halogen donor, it was really Golde and Thrush. And at AVCO, our initial question on that was, "Why is the spectrum where it is and not so broad?" We went on to making lasers from there.

JOHN M.J. MADEY

The Free-Electron Laser

John M.J. Madey received bachelor's and master's degrees from the California Institute of Technology in 1964 and 1965, and a doctorate in physics from Stanford University in 1970. He has remained at Stanford ever since, devoting most of his time to theoretical and experimental studies of free-electron lasers.

He spent five years laying the groundwork for the first demonstration of a free-electron laser amplifier, on January 7, 1975, which amplified the beam from an external carbon-dioxide laser. His group later demonstrated the first free-electron laser oscillator, operating at 3.4 micrometers. He also was involved in the first demonstration of a free-electron laser driven by a storage ring at the University of Paris-South in Orsay.

Now a research professor of electrical engineering and high-energy physics at Stanford, Madey's current interests include short-wavelength free-electron lasers, and developing free-electron laser technology on a laboratory scale for uses in medicine and other research.

C. Breck Hitz conducted this interview on April 1, 1985, in Madey's office at Stanford University.

L&A: When you were in graduate school, quantum electronics was a brand new field. What led you to get involved with it?
Madey: Well, I took a Bachelor's in physics, but even before then I can remember really being fascinated by electron tubes and microwave devices. I guess I'd been into ham radio fairly deeply in high school. As an undergraduate at Cal Tech, I spent a couple of summers working in a high-energy physics group and getting some exposure both to

particle physics and accelerator physics. That was interesting, but by the time I had been through a couple of summers on that and was ready to pick a graduate school, I'd kind of decided that high-energy physics was on too large a scale for me. I decided to look for something that you would do in a normal size laboratory. This was 1964, and lasers were coming on pretty strong then, so I decided to study quantum electronics.

L&A: You were doing graduate work at Cal Tech?

Madey: Right. I ended up working for Amnon Yariv for a year there, and getting a Master's in quantum electronics. Then, for various reasons, I figured I had missed the wave in quantum electronics and that things were starting to get a little bit dull. Looking back, that was probably a mistake, but I had done some reading on thin films and low-temperature physics, and I decided that maybe that was really the place to make a career. So I signed up with Bill Fairbank at Stanford and worked for the next five years in low-temperature physics in his group. That's where I got my doctorate.

L&A: Did the idea of a free-electron laser come to you somehow while you were doing low-temperature physics?

Madey: Actually, the idea came to me before that. I was taking Yariv's quantum electronics course at Cal Tech in 1965, and the thought occurred to me that the physics of the atomic or molecular laser was not really unique to an atomic or molecular transition. It seemed to me that if one had a free-electron transition of some kind, it could also — in principle, anyway — be capable of yielding gain. The example that came to mind in 1965 was Coulomb scattering from a nucleus, either in a plasma or solidstate, where you could generate a photon as a result of the collision.

The problem appealed to me because it involved a new laser theory. It was a new description of how one might be able to generate radiation, be it microwave or light radiation, from an electronbeam. And electronbeams, of course, were a long-standing interest of mine.

L&A: What was Yariv's reaction to the idea?

Madey: Well, Amnon may not remember it, but I remember it very clearly. We spoke after this particular lecture, and I said, "Look, suppose you had an electronbeam and you could scatter the electrons from a lattice, from the nuclei in a lattice, and you'd get some radiation in the final states. Isn't it true that you might be able to get some gain out of that?" And he said, "Oh, yes, that's right, that would

John M.J. Madey

work." And, in fact, he gave me a reference to one of his former colleagues at Bell Labs, Dietrich Marcuse, who had looked at that phenomena in maybe 1962 or '63, and used it as an example in his book. I looked it up and sure enough, there was gain, although not a lot of it. There were enough nonlinearities present to give it a fairly low saturation threshold. Besides that, it wasn't really clear that you could have a beam of free electrons running through a crystal without scattering from the phonons.

L&A: Should credit for the original concept of a free-electron laser should go to Dietrich Marcuse?

Madey: My guess is that, if you looked further, you would find others to whom the idea had occurred, and who had made at least some back-of-the-envelope calculations.

L&A: So one can't really say where the idea of the free-electron laser originated?

Madey: Well, the first reference that I've seen to the possibility, indeed the inevitability, of amplification in the scattering of free electrons was a 1964 paper by Dreicer at Los Alamos. He was looking at the attainment of equilibrium in plasmas. He pointed out that, if you just looked at spontaneous scattering events — in particular, at thermalization of electrons in an intense radiative environment — the cross-section for normal Thompson scattering would be too low. He realized you had to include stimulated radiation in the problem to get realistic thermalization times. And if you had stimulated radiation, you could get gain.

So, that was the first published reference I've seen to amplification from free-electron transitions. As I said, that was in the context of a plasma-physics problem.

L&A: You were saying that you left quantum electronics and came to Stanford to do low-temperature physics with William Fairbank. How did you wind up in laser physics again?

Madey: The discovery of the Gunn effect, in about 1966, perked my interest in free-electron lasers. You take a piece of bulk gallium arsenide, bias the sides relative to each other, use non-rectifying contacts, and you find that the thing oscillates. In fact, we get quite a lot of power out of those things; people don't make klystrons anymore. But it seemed to me, my God, this is really it! Here's your solidstate free-electron laser. This is really hot! If you recall, at first no one really knew what was actually going on in the Gunn effect. Nobody understood the microscopic physics, or even what was causing the oscillation. Now we know it's the formation of the low resistance domain as electrons scatter into a Gunn minimum, which then propagates across with some characteristic transition time and reforms. That's what's responsible for the oscillation. But at that time it seemed to me, gee, this might really be the microwave free-electron laser.

Well, the numbers didn't come out right when you tried to explain the Gunn effect that way. And then, within a relatively short time, someone provided the correct explanation. But the excitement had restored my enthusiasm to the point where I figured I'd keep kicking around the ideas. Bill Fairbank was very understanding about this. Even though I was working on a low-temperature physics problem, I kept stirring the thoughts on free-free interactions.

L&A: How long did this thought-stirring process last?

Madey: After one or two years, I came to understand more thoroughly why Coulomb scattering wouldn't work. Basically, the reason the gain was so low was that the transition probabilities for emission and absorption were almost identical. So there was very little net gain available. That's because the emission spectrum for Coulomb scattering is so broad. If you could get narrow-band emission from some sort of free-free scattering, your chances of gain would be a lot better. Once I realized that, I immediately thought of Compton scattering.

There was another important thing I'd found out about several years earlier. The Weizsacker-Williams approximation states that electrons moving at relativistic velocities cannot tell the difference between a traveling electromagnetic wave and a static electric or magnetic field. That is, even if you start with purely a magnetic field, the Lorentz transformation always generates an accompanying electric field in the electron rest frame.

Things really clicked at that point because now we could increase the amplitude of the scattering—use a superconducting magnet or a huge permanent magnet to create a much stronger equivalent field than you could get in a real electromagnetic wave. To me, it looked like we could probably get a reasonable amount of gain, even in the visible.

L&A: While you were figuring all this out, were you also working on your thesis in low-temperature physics?

Madey: Right. My thesis actually included both the low-temperature physics problem and the FEL theory. By 1969 I had a valid gain equation for an FEL and criteria for e-beam current, quality, and so forth. And I had some idea what we'd have to do to make an experiment work. In 1971, we started thinking seriously about building a free-electron laser. We had to generate strong, static, and periodic magnetic fields. And we had to have a good source of electrons at the proper energy, 20 to 50 MeV for emission in the infrared.

L&A: How long did it take to get it the first FEL to work?

Madey: We started research on the program in the fall of 1972, and that ran for four years before the first oscillator was demonstrated. But we demonstrated amplification, using a CO_2 laser as the driver, in 1975.

L&A: Did you have difficulty getting the work funded?

Madey: Well, those were the late '60s and early '70s, the golden years of physics research. We were running on about $200,000 a year from the Air Force Office of Scientific Research. That was quite an adequate budget in those days. In fact, we weren't sure we could spend it all when we started.

L&A: Did you encounter skepticism about the FEL from your colleagues?

Madey: Not at Stanford. We were fortunate that a new superconducting accelerator had just come on line here in the laboratory, and it was exactly what we needed. It had the right energy range, very low energy spread, excellent beam quality, and just enough current to bring the system above threshold. In fact, the laboratory staff and administration were fascinated about having an experiment which seemed uniquely suited to the capabilities of this new accelerator. So, we got good cooperation from the laboratory in setting this system up and in securing running time for it.

L&A: I recall a lot of people were running around saying that the FEL wouldn't work.

Madey: Ah...well, golly, I won't mention names, but I had people visit me who insisted that you would never see the narrow emission spectrum needed to make the laser work. Even after we had measured it, they insisted that it couldn't be correct.

But the important thing is that we had the commitment of the Air Force OSR, and they were willing to play out the string to see what we could do. We thought we understood the physics pretty well, you know.

L&A: Even so — were you surprised when the FEL first worked?

Madey: Nope. Things were very close to what we expected.

L&A: Were FELs being investigated anywhere else? Was there any concern that another group might operate an FEL first?

Madey: I think that, outside Stanford, there was, if not a wildly negative view, a sufficiently pessimistic evaluation of the possibilities, that we really didn't have to worry about competition.

L&A: Who contributed to the development of the FEL?

Madey: Three people in particular really deserve mention. One is Dick Pantell, who had done a calculation on the Compton scattering problem for free electrons in a real microwave field. He published his results with Soncini and Putoff in one of the IEEE journals in 1968. He had the right numbers for scattering from a microwave field.

The second fellow is Hans Motz, who's now at Oxford and was a research fellow at Stanford in the mid-1950s. Of course, he wasn't motivated by laser theory, because that hadn't been invented then. But he noted that if you ran an electronbeam through a periodic magnetic field, you would get a nice, narrow emission spectrum. He saw that, if you took advantage of the bunching that was inherent in the electronbeam from a linear accelerator, or if bunching could be induced in a traveling wave tube, you could get very high power output. Even if you got only spontaneous radiation, that looked interesting as a source. But, if you could get the electrons to radiate coherently in the undulator, it could really be a very intense source. So he built an undulator. He ran an e-beam through it, in maybe 1956, and saw this rainbow of colors. It was a really nice experiment.

The third individual is a chap by the name of R.M. Philips, who was working in GE's Palo Alto Research Laboratories in the late 1950s and early 1960s. He had gotten interested in the periodic-magnet geometry also. He made numerical calculations of the bunching caused by an input electromagnetic wave. He built an amplifier that ran at microwave frequencies and generated tens of kilowatts of average power at wavelengths around a centimeter or two.

L&A: Did Philips and Motz know about each other?

Madey: They may not have. I believe Philips was supported by a classified contract, so the full details of his research may not have been available in the open literature. I haven't seen cross references in their publications, and it may well be that they were working completely independently.

L&A: Were other people involved in the early work with FELs?

Madey: There was another analysis, one that's only recently become known, which I think is really remarkably prescient and complete.

It was prepared by Ken Robinson, who was an accelerator physicist at the Cambridge Electron Accelerator in Cambridge in the late '50s. He had developed a good classical model for the interaction, and he got the right numbers for gain. He prepared a rather detailed description of an infrared or visible oscillator and amplifier. The tragedy is that the Cambridge accelerator was closed down in 1966 or 1967. That accelerator was Robinson's baby, I guess. He was never professionally employed again. The analysis of the free-electron laser was discovered in a trunk of his possessions after his death.

L&A: It's a sad story.

Madey: It really is. The other tragedy is that, for its time, the Cambridge Electron Accelerator was a remarkably advanced machine. It might still be performing useful work today, certainly as a synchrotron radiation source, and perhaps even for FEL research. But apparently there was an agreement that called for the Department of Energy to return the site to its original state if funding was ever terminated. So the whole thing was knocked apart. Not a piece remains.

L&A: Well, that's really a discouraging story! What about the people who worked with you here at Stanford. Who were they and where are they now?

Madey: We had a remarkable group of people working here on that experiment. In particular, there's Luis Elias, who now heads up the FEL work at Santa Barbara; Dave Deacon, who still is pretty heavily involved here in short wavelength FEL developments; Todd Smith, who's continuing to push ahead in linear accelerator technology; and Bill Colson, who developed the classical theory for the FEL at Stanford, is still involved in the field as a consultant over in Berkeley.

L&A: Well, where do you see free-electron lasers going in the next decade or two?

Madey: It's a hard question to answer, particularly for large systems. FELs have benefited from a lot of enthusiasm over their potential, as well as from perhaps lack of full knowledge or understanding about the limitations and the technical difficulties. So, I have some concern about unanticipated problems that could slow down development of the big FELs.

As far as small FELs for scientific research are concerned, we've got a long-enough shopping list to keep the technology active and expanding for quite some time. One of the active programs here is the development of a small-scale infrared source for production of

A 10-meter free-electron laser amplifier at Stanford University.

*"Outside Stanford,
there was. . .if not a
wildly negative view,
a sufficiently
pessimistic evaluation
of the possibilities,
that we really didn't
have to worry about
the competition."*

modest power levels, ten to a hundred watts. Further in the future, we hope to put together an extreme ultraviolet source operable in the region below a thousand angstroms, perhaps to wavelengths as short as 200 angstroms.

L&A: What sort of power would you expect in the XUV?

Madey: It would be mirror-limited. If the only problem were the e-beam and the magnet, you might talk about powers of perhaps 100 watts to a kilowatt, from 1,000 angstroms to 200 angstroms. But I think mirror limitations will keep it far below that.

L&A: How far in the future is an XUV free-electron laser?

Madey: If we can get this thing started, we should have a system running in five years.

L&A: Would you do anything differently if you had it all to do over again?

Madey: I sure would. We spent a long time getting deeply into the laser physics from both the theoretical side and the experimental side. We kept the first experiment running for almost five years after the first demonstration because there were enough things that we didn't understand or wanted to get better diagnostics on. I think that was really a mistake. I think we made a mistake in not trying to develop less expensive ways of making FELs. We had an advanced super-conducting magnet and a superconducting accelerator, all of which were fine in terms of getting a clean beam and getting good data.

But, boy, we sure would have been better off if we'd stepped back and said, "Well, look, what could we do that would cost us, say, $100 a day to run instead of $10,000, even if it doesn't have some of these nice characteristics?" Because we knew enough at that time that we could have pushed towards such a system and had a small scale FEL operating before the end of the '80s. And FELs would be better off now if we had done that. In every successful area of laser technology, somebody's come along and figured out a cheap and dirty way to make the laser.

L&A: Like developing discharge-pumped excimers?

Madey: Exactly. And we put that off, I'm afraid, for five years longer than we should have.

BIBLIOGRAPHY OF LASER HISTORY

Mario Bertolotti, *Masers and Lasers: An Historical Approach* (Adam Hilger Ltd., Bristol, UK, 1983).

Nicolaas Bloembergen, "Nonlinear optics and spectroscopy," (Nobel Lecture), *Reviews of Modern Physics*, Vol. 54, No. 3, pp. 685-693, (1982).

Nicolaas Bloembergen, "Nonlinear optics, past, present, and future," *IEEE Journal of Quantum Electronics*, QE-20, pp. 556-558, (1984).

Nicolaas Bloembergen and Arthur L. Schawlow, "Bloembergen, Schawlow reminisce on early days of laser development," *Optics News*, Mar./Apr. 1983, pp. 13-16.

Joan Lisa Bromberg, "Research efforts that led to laser development," *Laser Focus/Electro-Optics*, Oct. 1984, pp. 58-60.

Joan Lisa Bromberg, "The construction of the laser," *Laser Topics* (Nov. 1985).

Sidney S. Charschan, "The evolution of laser machining and welding, with safety," *Proceedings of SPIE*, Vol. 229, pp. 144-153 (1980).

J.L. Emmett, W.F. Krupke, and J.I. Davis, "Laser R&D at the Lawrence Livermore National Laboratory for fusion and isotope separation applications," *IEEE Journal of Quantum Electronics*, QE-20, pp. 591-602 (1984).

Robert N. Hall, "Injection lasers," *IEEE Transactions on Electron Devices*, ED-23, pp. 700-704 (1976).

Jeff Hecht and Dick Teresi, "The short but tempestuous history of the laser," Ch. 4, pp. 49-61 in *Laser: Supertool of the 1980s*, (Ticknor & Fields, New York, 1982).

Jeff Hecht, *Beam Weapons: The Next Arms Race* (Plenum Press,

New York, 1984).

Willis E. Lamb Jr., "Physical concepts in the development of the maser and laser," in Behram Kursunoglu and Arnold Perlmutter eds., *Impact of Basic Research on Technology* (Plenum Press, New York, 1973).

Willis E. Lamb Jr., "Laser theory and Doppler effects," *IEEE Journal of Quantum Electronics*, QE-20, pp. 551-555 (1984).

Bela A. Lengyel, "Evolution of masers and lasers," *American Journal of Physics*, Vol. 34, pp. 903-913 (1966).

Eliot Marshall, "Gould advances inventor's claim on the laser," *Science 216*, pp. 392-395 (Apr 23, 1982).

Allen Maurer, "Masers to lasers, how they work and the men who made them," Ch. 3 of *Lasers: Light Wave of the Future* (Arco, New York, 1982)

S. Millman ed, *A History of Engineering and Science in the Bell System: Physical Sciences 1924-1980*; see Ch. 5, "Quantum electronics — the laser," pp. 151-210.

John W. Orton, D. H. Paxman, and J. C. Walling, *The Solid State Maser* (Pergamon Press, Oxford, 1970).

C. K. N. Patel, "Lasers— their development and applications at AT&T Bell Laboratories," *IEEE Journal of Quantum Electronics*, QE-20, pp. 561-576 (1984).

A. M. Prokhorov, "Quantum electronics," pp. 110-116 in *Nobel Lectures...Physics, 1963-1970* (Elsevier North-Holland, Amsterdam, 1972).

R. H. Rediker, I. Mengailis, and A. Mooradian, "Lasers, their development, and applications at MIT Lincoln Laboratory," *IEEE Journal of Quantum Electronics*, QE-20 pp. 602-615 (1984).

Arthur L. Schawlow, "Masers and lasers," *IEEE Transactions on Electron Devices*, ED-23, pp. 773-779 (1976).

Arthur L. Schawlow, "From maser to laser," in B. Kursunoglu and A. Perlmutter eds, *Impact of Basic Research on Technology* (Plenum Press, New York, 1973)

Arthur L. Schawlow, "Spectroscopy in a new light," (Nobel Lecture), *Reviews of Modern Physics*, V54 N3, pp. 697-707 (1982).

Arthur L. Schawlow, "Lasers in historical perspective," *IEEE Journal of Quantum Electronics*, QE-20, pp. 558-561 (1984).

G. F. Smith, "The early laser years at Hughes Aircraft Co.," *IEEE Journal of Quantum Electronics*, QE-20, pp. 577-584 (1984).

Howard M. Smith, "Historical introduction," Ch.1, pp. 3-12 of *Principles of Holography*, 2nd ed. (John Wiley & Sons, New York, 1975).

Peter P. Sorokin, "Contributions of IBM to laser science, 1960 to the present," *IEEE Journal of Quantum Electronics*, QE-20, pp. 585-591 (1984).

A. J. Torsiglieri and W. O. Baker, "The origins of the laser," (letter) *Science 199*, pp. 1022-1026 (1978).

Charles H. Townes, "Production of coherent radiation by atoms and molecules, Nobel Lecture" in *Nobel Lectures. . . Physics 1963-1970* (Elsevier-North Holland, Amsterdam, 1972).

Charles H. Townes, "Quantum electronics and surprise in the development of technology," *Science 159*, pp. 699-703 (1968).

Charles H. Townes, "The early days of laser research," *Laser Focus*, V14 N8, pp. 52-58 (Aug 1978).

Charles H. Townes, "Science, technology, and invention: their progress and interactions," *Proceedings of the National Academy of Sciences USA*, V80, pp. 7679-7683 (1983).

Charles H. Townes, "Ideas and stumbling blocks in quantum electronics," *IEEE Journal of Quantum Electronics*, QE-20, pp. 547-550 (1984).

Joseph Weber, *Masers: Selected Reprints with Editorial Comment*, (Gordon and Breach, New York, 1967)

Joseph Weber, *Lasers: Selected Reprints with Editorial Comment*, V2 (Gordon and Breach, New York, 1967)

Joseph Weber, "Evolution of the laser," *Federation Proceedings*, V24, Supplement 2-7.